The

Unaccountability

Machine

WHY BIG SYSTEMS MAKE
TERRIBLE DECISIONS—
AND HOW THE WORLD
LOST ITS MIND

DAN DAVIES

The University of Chicago Press

The University of Chicago Press, Chicago 60637
© 2024 by Dan Davies
Preface © 2025 by Dan Davies
All rights reserved. No part of this book may be used or reproduced
in any manner whatsoever without written permission, except in
the case of brief quotations in critical articles and reviews. For
more information, contact the University of Chicago Press,
1427 E. 60th St., Chicago, IL 60637.
Published 2025
Printed in the United States of America

34 33 32 31 30 29 28 27 26 25   1 2 3 4 5

ISBN-13: 978-0-226-84310-0 (cloth)
ISBN-13: 978-0-226-84308-7 (paper)
ISBN-13: 978-0-226-84309-4 (ebook)
DOI: https://doi.org/10.7208/chicago/9780226843094.001.0001

First published in Great Britain in 2024 by Profile Books Ltd.

Library of Congress Control Number: 2024949506

♾ This paper meets the requirements of ANSI/NISO Z39.48-1992
(Permanence of Paper).

To Tess, Joe, Poppy and Rosie

And also, to the middle managers of the world, the designers of spreadsheets and the writers of policies. Your work may be prosaic, but you are the ones who shape the world we live in.

Contents

Preface

One of my intellectual heroes, the Oklahoman 'cowboy philosopher' Will Rogers, once said, 'When you put down the good things you ought to have done, and leave out the bad ones you did do, that's memoirs.' Similarly, perhaps if I write down some of the smart things I should have included in this book and leave out all the dumb things I actually wrote, that might be a preface. Still, it's great to be able to sum up what I'm trying to say in the book, having had the opportunity to talk about it and argue about the underlying issues with some incredibly clever people since it was first published.

The Unaccountability Machine is about the biggest problem of modern industrial life – the problem of being overloaded with information, of 'trying to get a drink from a firehose'. There's a load of other stuff in the book – it's also a biography of an obscure 1970s scientist, a potted history of neoliberalism and an investigation into whether corporations are artificial intelligences – but these are all meant as examples and background explanations that demonstrate the big point I wanted to make: that we are all being overwhelmed by our mental environment, and this is causing problems. And specifically, problems of unaccountability.

I can't claim to be the first person to have noticed the problem; it's the subject of nearly every management book that's worth the paper it's written on. Ever since the birth of the large corporation in the twentieth century, people have been facing the problem that organisations get more complicated as they grow, and it's difficult to add enough capacity to manage this complexity. As a result, there's a cyclical process – things grow, become unmanageable and then they either reorganise or blow up. If they do the latter, there's either enough of an organisation left in the aftermath of the crisis to realise that change is necessary, or there isn't. And in such a way, history progresses.

When people are overwhelmed by information, they always react in the same way – by building systems. The system might be a simple rule of thumb, or it might be a computer program that needs to be run with 'big data'. It might be centralised and codified in a book of laws, or it might be distributed and emergent. What they all have in common is that they help to restore the possibility of decision-making, by narrowing down the space of things that need to be thought about.

Because they narrow the information space, every decision-making system is, implicitly, a model. We deal with the chaos of reality by constructing an imaginary world, making decisions about it and hoping they will work out in reality.

This is, broadly speaking, a good thing – after all, the ability to think in abstract terms is what sets us apart from other animals. Acting on the basis of consistent, simplified versions of reality is fundamental to our ability to work together, and we've been doing it for a long time. In a sense, even the Ten Commandments are a kind of decision-making system. The first nine commandments tell you what to do in common situations, whether on the Sabbath or when presented with the opportunity to commit adultery, while the final one goes further and tells you to not even think about coveting your neighbour's ox. As well as facilitating social life, such systems of rules free up mental energy for considering more important problems.

But scale matters, and the rapid onset of modernity has caused problems in this area – as well as in every other. In order to manage an industrial society, you need decision-making systems that work on an industrial scale, and we don't seem psychologically equipped to handle this.

'Accountability' is the word we use to describe the relationship between someone who makes a decision and its consequences. We've learned over time that when it's missing, not only do we tend to get worse decisions, but we also get worse people; being able to ignore our effect on the world seems corrosive to morality, and it makes people callous and capricious.

But in a space of time only slightly longer than a single human lifespan, we've created an entirely new kind of organisation. A very important consequence of industrialisation is that it breaks the connection between the worker and the product. A machine operator can't point to something and say 'I made that' in the way that a blacksmith can. So when our managers, administrators and bureaucrats are converted into operators of an industrialised decision-making machine, why would we expect them to feel any personal responsibility for the machine's output?

In principle, people can still overrule the systems in the interests of fairness – but the incentives all go in the other direction. Our organisations are set up to deliver lots of decisions, quickly, cheaply and with reasonable quality – on average. But that's the problem; you can't be fair to an average. You have to deal with individual people. So we have a world in which almost everyone is quite frequently subject to stupid decisions that they can't appeal against. And where the person who seems to be responsible denies all accountability and tries to blame it on a weird corporate entity.

This is why I think things really *are* getting worse – it's not just the natural perception of a fifty-year-old man that the world was a better place when my hair was thicker and my waist thinner. Everything got more complicated, and the only way to deal with that was to start weakening the failsafe concepts of accountability that protected us from bad decisions and bad people. And so, unless we find a new way to make decisions at scale, we're in trouble.

Can that be done? Will technology help? And what does any of this have to do with David Bowie or the invention of the automated gunsight? These are among the questions I'm trying to answer in this book – I hope you enjoy it.

Dan Davies, Exeter, December 2024

Author's Note

A great deal of intellectual energy is wasted on trying to attribute events to the categories of 'conspiracy or cock-up', when most of them should probably be blamed on something more abstract. History is the study of decisions, not of events, and many decisions are best understood as the outcome of larger systems rather than individual acts of will.

Systems don't have motivations, so they don't have hidden motivations. If the system consistently produces a particular outcome, then that's its purpose. But on the other hand, systems don't make mistakes. Just as it's impossible to get lost if you don't know where you're going, a decision-making system does what it does and then either lives with the consequences or dies of them.

This book talks about a lot of people, and at various points it might look like I'm criticising someone, often quite harshly. That's usually not what I'm trying to do. The people are brought into the story because that's the only way to sensibly describe the systems they were part of. If it hadn't been them, it would have been someone else. If this book had a real villain, we might identify it as 'the regrettable tendency of complex systems to have opaque and volatile dynamics'.

PART ONE

THE NATURE OF THE CRISIS

Introduction

> What counts as a crisis is the expectation of loss of
> control; in other words, cybernetic breakdown in an
> institution.
>
> Stafford Beer, *Brain of the Firm* (2nd edn), 1981

When we avoid making a decision, what happens to it?

This is a book about the industrialisation of decision-making
– the methods by which, over the last century, the developed
world has arranged its society and economy so that important institutions are run by processes and systems, operating
on standardised sets of information, rather than by individual human beings reacting to individual circumstances. This
has led to a fundamental change in the relationship between
decision makers and those affected by their decisions, the vast
population of what might be called 'the decided-upon'. That
relationship used to be what we called 'accountability', and this
book is about the ways in which accountability has atrophied.

In 1954, for example, Sir Thomas Dugdale resigned from
the British cabinet over something thereafter known as 'the
Crichel Down affair' (a relatively trivial piece of malpractice
relating to some farmland which had been requisitioned during
the Second World War). It was unclear whether he had actually been involved in the decision, but in the climate of the day

his position was untenable – as the minister in charge, he bore responsibility for what happened. This small but honourable act ensured that Dugdale's name will live on eternally; both he and the small area of Dorset which brought about his downfall are mentioned every time a modern British politician dodges the blame.

On the other side of the Atlantic, in 1953, President Harry S. Truman had a sign on his desk reading 'The Buck Stops Here'. Sixty-six years later, President Donald Trump mused that 'the buck stops with everybody' when asked who bore responsibility for a government shutdown. Anyone who has dealt with a corporation or bureaucracy of any size, and who is over the age of forty, is likely to have a vague sense that you used to be able to speak to a person and get things done; the world wasn't always a maze of options menus.

Industrial societies have been given a great deal of warning that something like this was going to happen. In the 1940s, the 'managerial revolution' was regarded as the likely outcome of the increasing complexity of the economic system and the scale of private and public-sector organisations which had developed to deal with it. The decline in individual accountability for unpopular decisions is not – or not only – a form of moral decline on the part of our rulers. It's also a consequence of the fact that there are fewer decision *makers* than there used to be. Nearly all the commands and constraints which afflict the modern individual, the decisions which used to be made by identifiable rulers and bosses, are now the result of systems and processes.

It is most obvious that something has changed when a decision which used to be made by a human being is made by a literal algorithm – a computer program. For example, people worry a great deal about, and legislators attempt to regulate, the practice of using artificial intelligence to decide whether you are turned down for a loan or denied insurance coverage. But a decision-making system is not just a computer. Financial and other decisions have been made on the basis of handbooks and lists of criteria since the days in which they were recorded with quill pens. The very parliamentary committees that attempt to regulate artificial intelligence are themselves bound by opaque and complicated rule books, designed to standardise their proceedings and to avoid having decisions attributed to any individual human being.

For a while, in the early days of this profound social change, people tried to understand what was going on. The study of decision-making systems was a big thing – they called it 'cybernetics'. But for a number of reasons, it never took off.

The most important reason was that the new science was a victim of its own technological success. The key figures in the early days of cybernetics – people like Norbert Wiener, John von Neumann and Claude Shannon – are almost all much more famous as pioneers of computing. In trying to invent a mathematical language to describe their problem, they quickly made a lot of discoveries relating to the representation of information and the operations that could be performed on it. These discoveries turned out to be extremely useful in the design of electronic circuits. Consequently, there was a huge and rapid brain drain away from the abstract study of decision-making systems and into the new industry of manufacturing computers. In many ways, when presented with such a golden opportunity, you would have had to have been rather odd not to have taken it.

Related to this fact, many of the people who were involved in the early days of cybernetics were a bit odd. In this book, we'll look at Stafford Beer, a British management scientist and consultant. Among other things in his life, he designed a method of holding conferences based on an icosahedron, and he tried to use an algae-filled pond as a computer. And he was the sensible one – the cybernetician who obtained high-powered consultancy jobs in industry and provided huge amounts of support to the *real* oddballs working in the field. The more I have looked into the science of 'management cybernetics', the more I have realised that there is no way to rid it of a fundamental core of eccentricity.

But something else also happened, because cybernetics didn't die a natural death. The most obvious place for these ideas about the organisation of a modern economy would surely have been within the subject of economics, but they didn't end up there. Some elements of management cybernetics – even some of the ideas of Stafford Beer himself – live on in business schools, usually as personal idiosyncrasies of individual researchers. In the economics departments, however, they have their own theories.

The way in which the economics profession ignored most of the work done in information theory is striking. It ignored its own traditions relating to uncertainty and decision-making, instead ploughing ahead with a frighteningly simplistic view of the world in which everything could be reduced to a single goal of shareholder value maximisation.

Why did they do this? Any account of how we got to where we are must cover three intellectual and organisational revolutions. There was one that happened but we forgot about it – the managerial revolution, where control was transferred from owners and capitalists to professional administrators. There was one that looked like it was going somewhere interesting but never arrived – the cybernetic revolution, which might have helped us understand that what mattered were the systems of professional management rather than the individual managers. And then, sometime in the early 1970s, we had the neoliberal revolution, which succeeded in shaping our society and its discontents, right up to the present day.

In a more fortunate counterfactual history, this book might have told the story of an academic debate between the neoliberal economists and the management cyberneticians, on the nature of a corporation and the organisation of decision-making. But that debate didn't happen. Or rather, it did happen, but not with diagrams and equations. It played out, very suddenly, with helicopters and guns at the presidential palace of La Moneda, in Chile on 11 September 1973, when the government of Salvador Allende was brought down by a coup led by General Augusto Pinochet, and Stafford Beer's consultancy project to bring a new form of economic organisation to the country came to an abrupt end.

Cybernetics was arguably in trouble long before the coup, but that was the moment that the world finally chose a different path. In doing so, we began to get into the habit of ignoring the fact that every year, more of the decisions that affect our lives are made not by people but by systems. Strange, alien intelligences with desires and drives quite different from our own. They're taking over the world – and not only that, for it seems that some of them are going mad.

But we should start this story at the beginning.

1

'Something's Up'

Capitalism is disappearing, but Socialism is not replacing
it. What is now arising is a new kind of planned,
centralised society which will be neither capitalist nor, in
any accepted sense of the word, democratic. The rulers
of this new society will be the people who effectively
control the means of production: that is, business
executives, technicians, bureaucrats and soldiers.

George Orwell, 'James Burnham and the Managerial
Revolution', 1946

'Yes. But those clowns put us in an awkward place where we're
going to have to thread the needle.'

An executive at Fox News called Ron Mitchell wrote those
words, in an email which later showed up in court documents,
and you could hardly ask for a better summary of the state of
Republican politics at the beginning of the 2020s. The litigation in which Mitchell was quoted related to a defamation case
which ended up being settled for $787.5m, the second largest
such case in American history

The nature of the defamation was about as serious as they
get; over a period of months, Fox News had repeatedly accused
Dominion Systems, a manufacturer of voting machines, of
conspiring to rig the 2020 US presidential election. This was,

to be clear, unequivocally false. Strangely, though, the emails and texts turned up in pre-trial disclosure seemed to confirm that nearly all the executives and journalists at Fox News knew that this was false, and that the people making the accusations were 'kooks', 'wackadoodle' and 'cruel and reckless'. But they kept broadcasting them anyway.

Why did they do it? This question never got put to them properly, because the case was settled on the courthouse steps. But the cache of emails suggests that they felt they had no choice. Having initially reported, correctly, that Donald Trump had lost the election, Fox News was completely unprepared for the consequent backlash from its viewers. The network had spent the previous four years working its audience up to a frenzy and the juggernaut couldn't be turned round – every attempt to break the bad news was treated as disrespect, and resulted in the loss of millions of viewers to less scrupulous broadcasters. In the words of one of Fox News's managing editors, 'weak ratings make good journalists do bad things'.

Many people destroyed their reputations and careers over the 2020 election, and the USA came considerably closer to an actual coup than anyone would have liked to have seen in the Capitol riots of 6 January 2021. The Fox News–Dominion litigation gives a window into what was happening, but it's a darkened window, through which it's difficult to really see the structure of cause and effect. Nearly all the individual people involved wanted to be doing almost anything else rather than what they actually did. Even when they were broadcasting things they must have known to be false, they hoped that they weren't believed. What was going on here? How did it come to this?

Crimes without criminals

I should probably introduce myself at this point. I have a professional interest in situations like this – cases where the exact thing that nobody wanted to happen has happened.

I was once an investment banker, and during the bouncing-rubble period of the last global financial crisis, I wrote a book called *Lying for Money*, about the history and economics of financial fraud. One of the big questions people were asking at that time was why, despite there being a strong popular perception that hanging was too good for them, bankers were not going to jail. I ended up reaching a pretty surprising conclusion: that although there were a number of bad reasons why prosecutors were reluctant to take on the vested financial interests, the basic problem was that in a democratic society, if you want to put someone in prison then you need them to intentionally carry out a specific criminal act.

Many years ago, in an earlier financial crisis linked to small savings banks in the USA during the 1980s, the fraud investigator Bill Black had coined the concept of a 'criminogenic organisation'. It referred to an institution in which incentives and management systems were structurally designed to ensure that crimes would be committed. This was an integral part of his theory of the savings bank crisis, that of a 'control fraud'. This occurred when a bank came under the control of a criminal, who was then able to set in place a system of incentives which ensured that fraudulent loans would be made and profits declared, which would naturally flow into the fraudster's pockets.

This solved several of the key problems that had bedevilled white-collar criminals over the years, the biggest of which was the need for overt acts of embezzlement. Stealing money by dishonesty had previously tended to require you to make illegal transactions – improper payments or false bookkeeping entries – which posed a risk of detection. In a control fraud, the dishonesty resided in the overall scheme and all the transfers of cash were legitimate – dividends, salaries, stock options and commercial transactions with parties connected to the management. Furthermore, there was often a number of layers between the guilty mind and the criminal act – unless there was a paper trail, most of the prosecutable things were done by people other than those that the authorities would naturally want to prosecute.

As it happened, the technology was in its infancy in the 1980s and most of the crooks pursued by Bill Black were careless about leaving the kind of paper trail he needed. But by the beginning of this century things had got weirder. After investigating some of the biggest scandals of the financial crisis, I concluded that not only was there no paper trail between the low-level crooks and the bosses, in most cases there was *actually no connection*. I ended up coining the term 'self-organising control fraud'. The idea was that the financial system of the developed world, from around the fall of Communism in Europe, had reached a point where the overall system of incentives in the economy was so criminogenic that banks had a natural tendency to organise themselves into fraudulent behaviour. All the top executives had to do was set unrealistic profitability targets and underinvest in legal departments and compliance systems. It wasn't so much that anyone had told their traders what to do; more that nobody ever organised things in such a way that they *wouldn't* form a criminal conspiracy.

The polycrisis and me

There is a definite family resemblance between the self-organising criminogenicity of the 2008 financial crisis and the stuff that went on at Fox News in 2020 – the 'criminogenic organisation' might really just be a special case of the 'structurally bad decision-making organisation'. But, when provided with a lot of spare time during the pandemic, I started wondering more about the raw material of the Capitol Hill riots. These angry people, whose reactions were so frightening, so difficult to control – where did they come from?

To a certain extent, of course, Fox News made them. Over twenty-five years, it had become a media colossus, largely based on two creative insights. First, that some people enjoy being angry, and will consume their rage as a product. And second, that some presenters have the gift of making people angry, just like a comedian can make people laugh.

But there's more to it than that. Fox News couldn't actually control the rage of its audience, as it belatedly found out. And it's not intellectually satisfying to create a theory of Trump voters that only explains that one phenomenon. Because almost everywhere you look in the world over the last couple of decades, you will find popular movements and trends that look quite similar. Silvio Berlusconi was the Trump of Italy. Jair Bolsonaro was the Trump of Brazil. Angry people, politically incoherent populist politicians and media organisations that sell productised rage are everywhere. It can't be a coincidence; we need a theory that covers them all.

The historian and writer Adam Tooze refers to a 'polycrisis' of the twenty-first century. This is an umbrella term covering a lot of these phenomena, held together by a kind of structural resemblance. The defining characteristic of the polycrisis is a loss of control by the previously existing hierarchy. It's often associated with an economic crisis of some sort, as in Greece in the early 2010s, but not always. There's a sudden loss of prestige, as the professional and managerial class makes a high-profile mistake. And the populists and the merchants of rage step into that gap. So the questions that need to be answered are: how does this happen? And why?

The people who left and came back

One of the stylised facts that interested me is that the people most involved in the ground level of the polycrisis were often the ones who had been disenfranchised for long periods before it began. You could see this in the Brexit vote in the UK. The turnout for that referendum was shockingly high, compared to the previous decade of general elections. The winning vote came from people who had not voted for a long time.

But why would they have bothered voting? Think about the period from the fall of the Berlin Wall in 1989 to the bankruptcy of Lehman Brothers in 2008. There had been hardly any choices to make. If you wanted to vote against globalisation, you couldn't. Wanted to vote for higher top rates of income tax or tighter bank regulation? Couldn't. Or the level of interest rates – it wasn't even possible to have a political opinion about it. You might be able to vote against further privatisation, but there was no political party that would reverse the old ones. If you were against immigration, you had no party to vote for either. All you had were choices between moderate centrist technocrats, competing on the grounds of who might manage the system more competently, in an economic context of gently growing prosperity. People lost interest because it was rational for them to lose interest – nothing needed to change and there was no way to change it.

Then the 2008 financial crisis happened, followed by a long period of recession and austerity, and suddenly it turned out that the technocratic consensus wasn't as competent or moderate as it had appeared. Ten to twenty per cent of the electorate suddenly realised that they might have to take an interest in politics after all. So they started paying attention again, and they didn't have the basic assumptions of the mainstream. All they knew was that the people who used to be in charge seemed to have screwed things up mightily.

The accountability sinks

Bad things were happening, but they didn't seem to be anybody's fault. This was infuriating to everyone; the CEOs, political leaders and other figureheads had been drawing big salaries and bonuses, and telling all of us that they knew best, but it turned out that they didn't even know what was going on in their own organisations. There was no mechanism to punish individual human beings, so the hoi polloi defaulted to what seemed like the only alternative – to use the democratic powers available to them to tear down the whole system.

The relationship between experts, decision makers and the general public had become completely dysfunctional. This wasn't really a crisis of managerialism or a crisis of political legitimacy – it was a crisis of accountability. And if accountability was at the root of the crisis, then maybe the things to look at were the mechanisms that cause it to be diminished.

Consider, for example, the following situation. A characteristically modern form of social interaction, familiar from the rail and air travel industries, has become ubiquitous with the development of the call centre. Someone – an airline gate attendant, for example – tells you some bad news; perhaps you've been bumped from the flight in favour of someone with more frequent flyer points. You start to complain and point out how much you paid for your ticket, but you're brought up short by the undeniable fact that the gate attendant can't do anything about it. You ask to speak to someone who can do something about it, but you're told that's not company policy.

The unsettling thing about this conversation is that you progressively realise that the human being you are speaking to is only allowed to follow a set of processes and rules that pass on decisions made at a higher level of the corporate hierarchy. It's often a frustrating experience; you want to get angry, but you can't really blame the person you're talking to. Somehow, the airline has constructed a state of affairs where it can speak to you with the anonymous voice of an amorphous corporation, but you have to talk back to it as if it were a person like yourself.

Bad people react to this by getting angry at the gate attendant; good people walk away stewing with thwarted rage, and they may give some lacerating feedback online. Meanwhile, the managers who made the decision to prioritise Gold Elite members are able to maximise shareholder value without any distractions from the consequences of their actions. They have constructed an *accountability sink* to absorb unwanted negative emotion.

The fundamental law of accountability

It is important to be clear, at this stage, exactly what an accountability sink is, and how they are constructed.* It's not

---

* If nothing else, you'll have a few tips about how to set things up in your own job to divert any troublesome accountability that might be building up.

just the way in which the hourly paid worker has been set up to act as a human shield. In order to make the sink effective, you need a *combination* of things: that person, plus a policy that there's no way to appeal the decision by communicating with a higher level of management. (Even if you somehow managed to get the CEO's phone number, you would come up against the fact that the policy was in place precisely to protect them from making that decision personally.)

So the crucial thing at work here seems to be the delegation of the decision to a rule book, removing the human from the process and thereby severing the connection that's needed in order for the concept of accountability to make sense. You could even coin a sort of law of management here:

> The fundamental law of accountability: the extent to which you are able to change a decision is precisely the extent to which you can be accountable for it, and vice versa.

The construction of accountability sinks has damaging implications for the flow of information. For an accountability sink to function, it has to break a link; it has to prevent the feedback of the person affected by the decision from affecting the operation of the system. The decision has to be fully determined by the policy, which means that it cannot be altered by any information that wasn't anticipated. If somebody can override the accountability sink and overrule a policy that is in danger of generating a ridiculous or disgusting outcome, then that person is potentially accountable for the outcome.

When an organisation decides to create an accountability sink, it's taking a risk. Implicitly, every rule is a model of the world – you can see both a model and a rule book as a relationship between causes and effects, inputs and outputs. But because they can't be a model of the *whole* world, both a rule book and a model have to leave out a lot of detail. That means that, like a model, a rule has to be based on assumptions about the kinds of situations that might need to have decisions made about them. So a notice on the wall of an office that says 'STAFF HAVE NO ACCESS TO THE SAFE' implies a set of assumptions about the kind of person who might walk in off the street and ask staff to open the safe, while a notice that says 'NO REFUNDS – NO EXCEPTIONS' carries an implicit assumption that there will never be a case where a customer is entitled to a refund.

When an unanticipated situation arises – either because something unusual has happened, or because the accountability sink was badly designed – there will be a mismatch between the input that the system anticipated, and what it actually got. And, because the system has been designed to work as an accountability sink, the outcome could be gruesome or absurd. Consider, for example, the tale of an airline and a few hundred furry mammals.

The squirrel shredders of Schiphol

Back in the 1990s, ground squirrels were briefly fashionable pets, but their popularity came to an abrupt end after an incident at Schiphol Airport on the outskirts of Amsterdam. In April 1999, a cargo of 440 of the rodents arrived on a KLM flight from Beijing, without the necessary import papers. Because of this, they could not be forwarded on to the customer in Athens. But nobody was able to correct the error and send them back either. What could be done with them?

It's hard to think there wasn't a better solution than the one that was carried out; faced with the paperwork issue, airport staff threw all 440 squirrels into an industrial shredder, apart from a few that had previously escaped from the animal containment facility. In later weeks, it transpired that this shredder was a specialised piece of machinery used in the poultry industry to dispose of worthless male chicks. Before the story hit the headlines, ground staff had over a period of months shredded 200 other squirrels, several dozen water turtles and a small flock of parakeets.

The press release in which KLM apologised for this horrible fiasco was a masterpiece of the genre that's still studied in business schools as an effective example of crisis PR. But it's less fascinating for its deflection of blame than for the underlying system that it revealed.

It turned out that the order to destroy the squirrels had come from the Dutch government's Department of Agriculture, Environment Management and Fishing. However, KLM's management, with the benefit of hindsight, said that 'this order, in this form and without feasible alternatives,* was unethical'. The employees had acted 'formally correctly' by obeying the order, but KLM acknowledged that they had made an 'assessment mistake' in doing so. The company's board expressed 'sincere regret' for the way things had turned out, and there's no reason to doubt their sincerity.

So what went wrong, and who was responsible for shredding the squirrels? The first question is easier to answer than the second. KLM had set up a system whereby decisions about animals with the wrong import paperwork were left to someone

---

* The De Meern Foundation for the Shelter of Squirrels, the Netherlands' only specialist squirrel rescue organisation, was particularly annoyed about not having been asked whether it could help.

at the agriculture department. In doing so, everyone involved had accepted that a low baseline level of animal destruction was tolerable – which is why they bought the poultry shredder. But, in so far as it is possible to reconstruct the reasoning, it was presumed that the destruction of living creatures would be rare, more used as a threat to encourage people to take care over their paperwork rather than something that would happen to hundreds of significantly larger mammals than the newborn chicks for which the shredder had been designed.

The characterisation of the employees' decision as an 'assessment mistake' is revealing; in retrospect, the only safeguard in this system was the nebulous expectation that the people tasked with disposing of the animals might decide to disobey direct instructions if the consequences of following them looked sufficiently grotesque. It's doubtful whether it had ever been communicated to them that they were meant to be second-guessing their instructions on ethical grounds; most of the time, people who work in sheds aren't given the authority to overrule the government. In any case, it is neither psychologically plausible nor managerially realistic to expect someone to follow orders 99 per cent of the time and then suddenly act independently on the hundredth instance.

For their part, the people in the agriculture ministry were a long way from the workers who had to carry out their instructions. They had a responsibility to protect the biosecurity of the Netherlands and implement the relevant European regulations that required health checks for imported squirrels. A policy stating that 'commercial imports of pets must provide veterinary paperwork, or else the animals will be returned to their port of origin or euthanised at the expense of the importer' looks sensible enough, and would cover the overwhelming majority of cases. And with the policy implemented, each decision to order the destruction of animal cargo is just a matter of following the policy.

That's the purpose of making policies – to reduce the amount of time and effort spent making decisions on individual cases. However, it's also the root cause of this sort of problem. When you set a general policy, you either need to build in a system for making exceptions (and make sure that it is used), or you need to be confident that all the outcomes of enforcing that policy will be acceptable. In the case of the policy with regard to rodent imports, this wasn't the case; the Schiphol incident led to an emergency debate in the Dutch parliament. The top management of KLM were, in fact, so appalled that within a few months the airline had a new policy of refusing to ship any exotic animals at all – while Schiphol Airport came close to losing its certification as an import centre because the company that ran the 'animal hotel' there had gone bust.

Although dramatised by grim humour, the KLM squirrel debacle illustrates a few potentially important things about the underlying reality of management and information. A decision with no real owner had been created because it was the outcome of a process. The process worked well, until something that hadn't been anticipated (the pet squirrel craze)· showed up, and then it delivered disastrous results. There was no effective way in which information could be fed back to the people who could change the policy, so the decisions continued to get worse. And then, when something so outrageous happened that it couldn't be kept out of the newspapers, there was nobody to blame.

This property of there being 'nobody to blame' is the definition of what constitutes an accountability sink. It's not clear what KLM should have done when faced with a consignment of 400 squirrels from a breeder who refused to obey the import regulations. The best solution would have been to refuse to load the cargo in Beijing, but the plane had already flown. Tragically, the decision to put them down and then bear the public opprobrium might even have been correct. But making a specific decision to kill the squirrels would have been much less psychologically tolerable for the policy-making managers than simply creating a system which ensured that they would be shredded unless a lowly employee disobeyed a specific order. In some ways, a disaster like the Schiphol squirrel episode can be seen as the policy mechanism providing one of its intended functions – acting like a car's crumple-zone to shield any individual manager from a disastrous decision.

Accountability and its discontents

There are a number of reasons why people might construct accountability sinks. The most fundamental one is that being held accountable for things is horrible. The key privilege of being a manager, surely, ought to be that you have the status of *being a manager* rather than one of the managed. Having your decisions questioned, and having pressure placed on you to change them – which, if the fundamental law above is correct, is the essence of accountability – is humiliating and unpleasant. Not only that, but in large organisations, the kind of conflict that's implicit in a system where individuals make decisions is potentially corrosive of trust and relationships. Let's consider, as an example, the accountability sink in the academic publishing industry, through which scholarly papers are gathered in journals and sold to university libraries.

Once upon a time, academic publishing was largely a non-profit affair – most scientific and humanities journals were published either by universities or by learned societies, to keep their members up to date with new research. The articles were selected by editors and put through a system of 'peer review' whereby journal publishers would request comments from academics who were specialists in the field (anonymously, to ensure that the reviewers were objective and so people would not fear the professional consequences of criticising a fellow scholar). On the basis of this review, papers would be revised, resubmitted and accepted for publication. It was (and is) a laborious process, and many owners of journals were glad to sell them to private-sector publishers.

Academic publishing is extremely profitable. The foundations of Robert Maxwell's media empire were built on Pergamon Press, which was one of the first companies to realise that the business model has two incredibly favourable properties. Firstly, the customer base is captive and highly vulnerable to price gouging. A university library *has* to have access to the best journals, without which the members of the university can't keep up with their field or do their own research.

Secondly, although the publishers who bought the titles took over the responsibility for their administration and distribution, this is a small part of the effort involved in producing an academic journal, compared to the actual work of writing the articles and peer-reviewing them. This service is provided to the publishers by academics, for free or for a nominal payment (often paid in books or subscriptions to journals). So not only does the industry have both a captive customer base and a captive source of free labour, these two commercial assets are for the most part the same group of people.

A not-wholly-unfair analysis of academic publishing would be that it is an industry in which academics compete against one another for the privilege of providing free labour for a profit-making company, which then sells the results back to them at monopoly prices. It is, as you'd expect from that description, highly profitable – and passionately hated by the academics.

However, the model persists because somewhere along the way, the journal industry managed to insert itself into the staff promotion and recruitment function of universities all over the world. In doing so, it created an extremely useful accountability sink for senior academics and managers of universities, while also solving an awkward and unpleasant interpersonal problem for them – how to judge the quality of scholarship without offending the scholars.

The truly valuable output of the academic publishing industry is not journals, but citations. Academic papers cite one another, and the best ones get cited a lot. Some journals tend to systematically publish more of the highly cited papers than others, and so these are regarded as the best journals with the highest standards. If you can work out which are the best journals, and which scholars publish in them, and which papers get the most prestigious citations, then you can use fairly standard statistical techniques to generate a 'score' for every academic, describing how well-cited their work is compared to their peers. The process is quite similar to the PageRank algorithm used by Google to decide which web pages to show first in search results.

At this stage you might suggest that 'indexing billions of web pages' and 'assessing the influence and quality of scholarship' are very different tasks; surely nobody should expect an algorithm designed for one to be any good at the other. Set against that objection is the fact that if you have to decide which academics should be promoted or employed, the 'weighted citation count' is a perfect accountability sink. Academic politics is notoriously vicious, and academic careers tend to intersect a lot – what goes around comes around, and people need to collaborate. In that sort of environment, a system in which academics directly assessed each other's promotion cases would cause all sorts of interpersonal problems; it would be difficult to work productively with someone if you were known to have previously judged their research to be less excellent than one of their peers.

So although the citation index is in all probability a bad measure that seems to lock the universities into an expensive and unsatisfactory publishing model, the outsourcing of the academic performance measurement system is a solution rather than a problem. It redirects potentially destructive negative emotions to a place where they can be relatively harmlessly dissipated.

How health and safety goes mad

The accountability sink doesn't just dissipate resentment and jealousy – deployed correctly, it can shield organisations against legal liability and similar threats. A decision made by an individual can be second-guessed and questioned, and it can be the object of litigation. A policy, on the other hand, is harder to challenge, particularly if it is made public. Fairly obviously, if decisions are taken in line with a pre-existing policy, they cannot be characterised as the result of caprice or prejudice. And if a policy is made publicly available ahead of time, anyone who interacts with the organisation can be argued to have implicitly accepted it. None of these defences are bulletproof – KLM's policy didn't save them from public shaming in the squirrel-shredding episode – but they provide significant shields against material business risks.

Accountability sinks are even more important when the insurance industry gets involved. Insurance is a business of risk pooling – it only works because the average of a lot of similar events is much more predictable than any single event. Actuaries (the specialist mathematicians who set pricing and terms for insurance policies) work by creating averages and probability distributions, not by analysing the unique and specific characteristics of every possible outcome someone might want to insure. For this reason, insurance underwriters love consistent processes; standardisation is a key part of what they do. Once a risk assessment is subject to a set of rules, the class of uncertain events covered by the (corporate) policy can be covered by an (insurance) policy.

When a risk is covered by an insurance policy, however, being accountable for a decision becomes not just psychologically unpleasant, but financially risky. If you've agreed with your insurer that something will be required or forbidden, then even if it seems obvious that an exception should be made, it's possible that you would be invalidating your policy to do so. If something goes wrong, the insurance company is often not set up to listen to your explanations; it has hundreds of customers, and it would be unrealistically expensive for it to employ loss adjusters to adjudicate the validity of every single claim.

How crises happen?

I have to admit, I got pretty excited about the accountability-sink concept when I came up with it, a few years ago. We were looking at a crisis of legitimacy, which involved a crisis of managerialism, and they were both really crises of accountability. The last few decades had seen the rise of the professional and managerial class in the economy and society. These people had been able to reorganise and re-engineer many of the most important institutions from politics to business to finance and the law – and they had done so in order to reduce the extent to which they could be held responsible for their actions.

They had done this partly because it was psychologically more pleasant for them to do so, and partly because they weren't a unified class – they were at risk from one another, and accountability sinks were a useful way of redirecting what might otherwise be sources of conflict. There was a sort of neatness to the concept – by reducing their ability to make decisions as individuals, the professional and managerial class cemented their control of the overall system. It all worked perfectly, except for when it didn't.

One thing made my idea attractive as a Big Theory of Everything: it had a built-in prediction of periodic crisis, to make it more interesting and to allow for some movement in the forecasts. Because the rules had to be inflexible in order to absorb personal accountability, they were unable to react to situations that hadn't been anticipated. It was to be expected that anomalies and inconsistencies would build up over time, eventually causing the system to fail catastrophically – rather like a self-driving car handing back control to the owner at the worst possible moment.

Can you blame me for feeling a little bit bumptious as I went around telling people I had figured it all out? There were various parallels to the work of some of my favourite authors. Nassim Nicholas Taleb had published *Skin in the Game* a few years earlier, emphasising the importance of robust decision rules based on individual personal accountability. The mechanism through which periods of calm resulted in inflexible structures that resulted in crisis was similar to Hyman P. Minsky's financial instability hypothesis, which held that the business cycle was driven by people's tendency to build up unsustainable debt burdens when economic conditions were favourable. And there was even a link to a theory that could explain why the crisis of managerialism had resulted in a crisis of systemic legitimacy.

The deep state

My favourite historian of conspiracy theories, Peter Dale Scott, had been using the concept of accountability reduction as a central organising principle of his work since the 1980s. Rather than 'conspiracy', he preferred to use the term 'parapolitics', defining that as the part of the government system in which the possibility of accountability is intentionally diminished. He later supplemented this with the idea of 'deep politics', referring to the political practices and arrangements which were kept out of public discourse, and in the 2000s, he began to refer to the 'deep state'. This phrase was originally used by political commentators in Turkey as a simple factual shorthand for the network of relationships between the army, media and large industrial companies in their country, but it caught on rapidly in the English-speaking world when Donald Trump began to use it to refer to anyone who stopped him from doing what he wanted.

It's unlikely that Trump had read much Scott, but there was clearly something that resonated with his supporters about the idea of a part of the state which operated at least partly by removing itself from accountability. And it's hard to say that they were completely irrational or conspiratorial in their worldview. When Peter Dale Scott began to write about parapolitics and the conscious reduction of accountability, he was mainly referring to the military and intelligence services. But if you look at how the world has developed over the last four decades, there's hardly any 'non-deep' state left.

For example, look how the political system has used 'the

market' as an accountability shield since the early 1980s. Whether in its guise as 'the bond market', which necessitates austerity, tax cuts and deregulation, or just as a generalised amorphous maker of decisions that can't be blamed on any individual, the market has encroached on areas which were previously considered the normal business of government. Nationalised industries were privatised, then often sold overseas. Even some fundamental functions that older political scientists might have seen as core to the nature of government, like the power to set the level of taxes or control of the money supply, have been to a greater or lesser extent removed from the political sphere.

The uncrowned fiscal king of Europe

In 1995, Europe was in the process of establishing a single common currency. One of the biggest obstacles was the concern from Germany and other wealthy countries that they might end up having to bail out weaker economies. In order to overcome this objection, the Euro member countries came to an agreement, known as the Stability and Growth Pact, that they would commit to cutting spending or raising taxes if the gap between the two (the 'fiscal deficit') grew too large when measured as a percentage of GDP.

But what should that percentage be? The compromise ended up being set at 3 per cent. This number had been used for the last decade in France, a budgetary rule of thumb to ensure that deficits didn't get out of control. The exact number had been suggested by a civil servant called Guy Abeille in 1981.

According to an interview given by Abeille towards the end of his career, he had been asked by the budget director for 'an easy rule, that sounds as if it comes from an economist, and can be opposed to the ministers that walk into my office asking for money'. At the time, the French deficit was a bit more than 2 per cent of GDP, so it made no sense to suggest anything lower; 3 per cent was 'a good number, a number that has gone through the ages, it made one think of the Trinity'. The process by which the attractiveness of various small numbers was discussed before 3 per cent was chosen apparently took less than an hour.

By 2010, the Stability and Growth Pact had resulted in a lost decade of growth for Italy, and very nearly saw Greece ejected from the Euro. Abeille had not known that he would be setting such an important policy for an entire continent, and he had certainly never been elected to the job. But the target became important to the entire European project precisely because, once enacted, it was an accountability sink, defusing the political conflict that would have been opened up if there had been a public debate in which German politicians discussed what social programmes and benefits needed to be cut in Greece and Portugal.

The Stability and Growth Pact was an accountability sink for the Euro – as demonstrated by the political conflict when it had to be abandoned in the face of financial crisis. And the decision to import the 3 per cent threshold from the French budgetary framework was an accountability sink for the Pact – to have carried out a detailed analysis of the subject would have involved someone deciding whether applying a single percentage for every country in the Euro zone made any sense. This sort of thing happens distressingly often in policy contexts. All over the world, the reason why most central banks have inflation targets of around 2 per cent per year is that way back in 1988, Roger Douglas, the finance minister of New Zealand, was asked on TV whether he was happy that inflation had just dipped below 10 per cent. Off the cuff, Douglas replied that he would rather see it below 1 per cent. Some statisticians at the RBNZ calculated that their measure of inflation probably had an upward bias of about a percentage point, and so Don Brash, the Governor, announced that the new policy framework would target a measured rate of 2 per cent. And afterwards nearly everyone who subsequently decided to copy the Kiwi policy framework copied this inflation target, too. If you trace back many important decisions of the last few decades, you will regularly come up against the uncomfortable sensation that the unacknowledged legislators are relatively junior civil servants who put placeholder numbers in spreadsheets, which are later adopted as fundamental constraints; to do otherwise would mean someone having to risk being criticised for making a decision.

IMF riots

There is only one situation in which the avoidance of account-ability is widely analysed by economists, and that is the case of the International Monetary Fund. The phenomenon of civil unrest taking place shortly after the agreement of an IMF or World Bank loan is common enough in the history of Latin America and Asia for the term 'IMF riots' to be a common subject of study in the development economics literature. As a concept, the IMF riot looks quite easy to explain; the loans tend to be agreed alongside 'structural adjustment programmes' which have often included quite severe budget austerity, the removal of subsidies on fuel and food, and other obviously unpopular measures. But the relationship is not necessarily one of simple coercion.

In 1998, in a review article on the development of its policy on conditionality, the IMF itself said that 'weak governments like to be able to reduce the domestic pressure applied by interest groups and political parties by pointing to the need to respond to an alternative pressure coming from the outside. In the course of the 1960s, the IMF became accustomed to being used in this way as an external whipping boy or scapegoat.' Entire journal articles have been written about 'the scapegoat function of the IMF' and how it might optimally be deployed by domestic elites in poor countries in order to impose policies that they agree with, but which they might not care to be associated with politically.

Something feels wrong about this, though – in specialist journals and internal discussions, the idea of tricking the public into blaming a global institution in order to solve a domestic political dilemma might seem quite clever, but few economists would care to defend it to a crowd of people gathered in a square because they do not have enough food to eat. Or even to defend it outright at a dinner party in a comfortable developed world country; on most of the occasions when I have referred to the 'scapegoating' literature on IMF riots to people who are economists, but not specialists in development, their initial reaction has been to assume that it was left-wing propaganda. Just as accountability is psychologically intolerable to the decision makers, its absence is intolerable to the subjects of their decisions.

The 'managerial revolution' happened relatively recently in historical time (the book of that name was written by James Burnham in 1941 and was intended to describe what Burnham saw as a new reality). The increasing domination of the industrial world by accountability sinks is even more recent. Humanity simply hasn't had enough time to get used to it. That's one big reason why we have IMF riots; it matters a lot to populations all over the world whether the policies they have to suffer under are being imposed on them by people and institutions they understand and feel they can communicate with, or by impersonal and abstract entities that have to be experienced as forces of nature.

For nearly all of history, there have been two kinds of authority taking the big decisions affecting people's lives. There is a fundamental distinction between 'kings' and 'priests'. A king might be more powerful, but his orders can be argued against – it might be inadvisable to do so, but if you can change the king's mind you can change the decision. The priest, on the other hand, gains his authority from his status as the interpreter of the Word of God, so his decisions are considerably more difficult to reverse. This means that it matters a great deal which kinds of decisions are given to which kinds of authorities, and the question of the boundary between the two spheres has often been one of the central issues of entire eras – it was the subject of the Thirty Years War in Europe. A lot of the discontent in the modern world might come from having taken decision-making structures that were designed with 'king-like' leaders in mind, and handing them over to managers who didn't act in the same way.

A premature solution

This was the nature of my theorising, as my mind wandered and I imagined being acknowledged as the guru who had identified the cause of the problems afflicting our society. As I typed up my thoughts, I dreamed of giving talks to huge audiences, gently chiding the managers of the world for avoiding accountability, and perhaps ushering in a new age of responsive government. I had even, to my utter shame, begun to coin a law that I could foresee appearing in books of aphorisms alongside Murphy's Law and the Peter Principle:

> The principle of diminishing accountability: Unless conscious steps are taken to prevent it from doing so, any organisation in a modern industrial society will tend to restructure itself so as to reduce the amount of personal responsibility attributable to its actions. This tendency will continue until crisis results.

I can still remember the crushing disappointment of realising that it was all much too simple.

It's more complicated than that

How much discretion does anyone really want individual middle managers to have? If the 'fundamental law' is correct to link accountability with the possibility to change things, then the only way to bring accountability back into the system is to increase the ability of individual human beings to make exceptions to policies. And that's often bad; many of these policies were brought in specifically because the individual human beings were being capricious, unfair, even racist. It's good that people can understand what's going to happen by consulting a rule book, rather than having to guess what somebody else is going to decide.

And are the accountability sinks really as inflexible as all that? It could equally be argued that in the Schiphol squirrels case, KLM changed their policy. It feels like a catastrophic failure because it was such an embarrassment, but there was a system there, and it had a fallback mode. Even the IMF riots could be seen as the ultimate appeal to higher authority – the political system can try to get rid of accountability, but when the consequences go beyond their tolerances, they're reminded that the system is dependent on the consent of the governed. Many of the things I've identified as 'accountability sinks' could just as easily be called 'the rule of law'.

The common law of England and Wales, considered as a decision-making system

And thinking about the law in this way suggested to me that there can be good and bad accountability sinks. Obviously, you don't want judges to be personally accountable in the sense of being able to decide what they want. To this extent, the law absolutely fits the definition of an accountability sink – it's a system which is meant to provide decisions that are not attributable to an identifiable person, and which are not alterable in response to feedback from those affected by them. You can't argue with a judge.

But that's not the end of the story; it's not a complete accountability sink and the feedback link isn't completely broken. The law is made up mainly of precedent rather than statute, and its development takes place by a kind of feedback mechanism. Precedents are more or less influential depending on the seniority and reputation of the judges who made those rulings, and the process for promoting and recognising judges does take into account the kinds of decisions they make over time.

Interestingly, this sort of feedback, with accountability for decisions happening slowly through the accumulation of reputation against named jurists, also happens in the theology of the major world religions. People who want to break the link to human decision makers and treat the books of law as a source of algorithmic judgement are called *fundamentalists*. Or, if it's the US Constitution that they're trying to pull this trick with, 'strict constructionists'. Either way, they're dangerous. It's probably indicative of the attractiveness of the accountability sink concept that people keep reinventing fundamentalism, even after it's failed so badly so many times.

I put my celebratory glass of wine down, half-finished, and looked at my bookshelves. A good friend had lent me some books several years earlier, which I had never opened because they looked difficult. He'd said at the time that there was only one really good treatment of the importance of information and feedback in the social sciences, and that it was a shame that it had never caught on. I reached up to the top shelf and pulled down the first volume that my hand landed on. It was called *Brain of the Firm*, by Stafford Beer.

2

Stafford Beer

Stafford talked and talked about the work his group was
doing with boundless enthusiasm and good humor. One
by one the visitors left until only David and I remained.
And still Stafford held forth about the future scope for
Operational Research, chain-smoking cigars and with
frequent recourse to a hip flask. When he left to go to his
room he still seemed fresh – but we were exhausted by
his sheer exuberance.

> Jonathan Rosenhead, 'Stafford Beer', in Assad and Gass
> (eds), *Profiles in Operations Research*, 2011

There's a fantastic old BBC TV mock-documentary called
Time on Our Hands, which is still available online. Made in
1963, it invited people like Aldous Huxley and Kingsley Amis
to speculate on what the world might look like if the most opti-
mistic predictions of technological revolution were to come
true – how would society have to change to deal with a world
of abundance and unlimited leisure? Twenty-five years later,
they revisited the predictions. Some of the contributors had
died in the meantime, but Stafford Beer was still there and the
contrast between the two is striking. In the archive footage
he's every inch the management guru – powerfully built in a
double-breasted suit, with a thick and neatly trimmed black

beard, holding a cigar and making forceful statements about the potential of electronic computing. In the studio discussion in 1988, he looks more like a guru of the literal kind – the beard is grey and flowing, the clothes hippy-ish and his comments full of rueful wisdom as he considers roads not taken.

Stafford Beer had been invited to that documentary because in 1963, that was the kind of status he had. He was perhaps not quite at the same level of general recognition as Amis and Huxley, but he was a well-known commentator on business affairs, and very much respected as a public intellectual. In particular, he was one of the country's most prominent experts on, and advocates of, the use of computers in industry. This was the year in which Prime Minister Harold Wilson made a speech about 'the white heat of the technological revolution', as a possible way for a post-imperial Britain to find a global role. It was quite natural for the broadcasters to turn to him, as a good public speaker and interesting thinker, and to be able to caption him as a 'Cybernetician' made the programme look new and futuristic. (The 'Cybermen' made their first appearance on the *Doctor Who* television show three years later.)

A lot of things happened to Beer over that twenty-five-year period. By 1988, his fame had very much shrunk. He was still well regarded among management theorists, but his career had reached the stage of holding down a portfolio of visiting professorships and chairing learned societies – one step away from 'emeritus'. He continued to present his ideas on the technological revolution and the need for social adaptation, but although people applauded, they didn't listen. By the time of his death in 2002, the obituaries needed to remind people who Beer was, and why, for a short 'cybernetic moment', it had been thought that he was inventing the future.

People used to get nervous about Stafford Beer. In an

academic world where specialists tend to be quite defensive about their fields of expertise, there was often a sense that he was practising without a licence. He had been a psychologist without a medical qualification, a mathematician without a maths degree, an economist without a PhD. He made his career in management consultancy, where the sense of outsiders advising birds on how to fly is palpable, and the lack of any sort of professional standards body an embarrassment.

And yet Beer worked in all of these fields, and people with those qualifications respected him. More than that – people with the right PhDs, professorships and even Nobel Prizes often looked up to him, cited his papers and asked him to be president of their societies. If he was a bluffer, he must have been the best bluffer of the twentieth century.

Or, possibly more likely, he was one of the great autodidacts; a natural mathematician and instinctively multidisciplinary thinker, who was diverted from a conventional academic path by the Second World War. He left a degree course in philosophy and psychology in 1944 to join the army, ending up in India as a captain in the British Army intelligence corps. After the war, he was given a job as an army psychologist and put in charge of a unit of illiterate soldiers who had been diagnosed as having 'psychopathological personalities'.

It was a different time as far as experimentation on human subjects was concerned. Beer's unit had never been given any form of medical treatment as there were no resources for it – and no effective treatment existed for whatever conditions they had. But the psychology professor from his half-completed university degree vouched for him and no one objected to the idea of allowing an unqualified but interested officer to have a go and see what might be achieved.

Beer approached the problem similarly to the challenges he

had faced as an intelligence officer, in keeping track of the shifting allegiances and balance of forces in the Indian subcontinent leading up to Partition. He made observations of the facts in front of him, deciding what variables to monitor and plotting their trends over time. Most importantly, he refused to make assumptions about what was going wrong in the heads of the young men in his charge.

The black boxes

Beer noticed that 'they could not write a letter home, nor read a newspaper, and such sums as $4 + 3 = ?$ often had them fooled. But they could debate with great energy and verbal facility if not felicity; they could play darts – "21 – that's 15 and a double 3 to go"; and they could state the winnings on a horse race involving place betting and accumulators with alacrity and accuracy, and apparently without working it out.'

These troops were for the most part happy as individuals and formed a unit that managed itself and maintained a form of coherence. Stafford Beer decided that this would be his starting point; rather than creating a theoretical framework, he worked iteratively, trying something and seeing what happened, before following feedback to do more of what worked and less of what didn't.

Although it seems obvious now, this was quite an unusual way to go about things in 1947. In some economics departments, it's almost heretical even now to make lawlike generalisations about human behaviour that aren't reducible to people trying to maximise their satisfaction according to some underlying and possibly unknown set of preferences. Later on, Beer would identify it as one of his fundamental axioms: 'It is not necessary to enter the black box to understand the nature of the function it performs.'

The underlying idea is somewhat stronger than this: if the black box is a complex system, it's likely to be pointless – or even dangerous – to try to understand its inner workings and use that understanding to manipulate a precise outcome. This is a matter of respecting the complexity of the problem – a genuinely complex system is one in which you cannot hope to get full or perfect information about the internal structure, and cannot have any acceptable degree of confidence that the bits of information you don't have can be safely ignored. Rather than trying to use a mixture of partial information, preconceived theory and guesswork, you need to step back, accept that the system will keep its secrets, and observe its behaviour.

And of course, this means that different observers might have different opinions. The property of 'being a black box' isn't an objective one – it's a description of a decision taken by someone working on the system that they don't have enough understanding to safely treat it in any other way. In some of Beer's writings, the property of complexity itself is also dealt with in this way; rather than trying to count combinations of connections, you define what it is for something to be a complex system by saying that it's one which has to be treated as a black box.*

In the twenty-first century, with the ideas of chaos theory, big data and opaque algorithms widely distributed, this might seem platitudinous, but it's not. The idea that systems can have different properties from the sums of their components, with properties that can't be deduced from those components, is

---

* If you can control something in your factory like a machine, it's a 'clear box' in this language; there are even 'muddy boxes', where more detailed analysis could help understand the relationship between inputs and outputs but the system keeps a few secrets because really detailed analysis would require disproportionate effort.

dangerous for many institutions. It's inimical to a particular model of scientific enquiry, one where understanding is built up from small and simple to big and complicated – appropriately, given Beer's background, it's more consistent with neurological and psychiatric approaches, where medical science still has surprisingly little idea of how its effective treatments might function.

Homo economicus and his friends

Beer didn't immediately realise how new this worldview was, but it turned out to be a quite important obstacle to communicating his ideas to the people who might otherwise have been his natural audience, economists. Economics was at this time reorganising itself as a science based on what might charitably be called 'an engineering model' and less charitably 'physics envy'. In the 1940s and 1950s, there was a methodological revolution in economics, as a series of new mathematical techniques were invented and put to use.

Some of these techniques were, indeed, carried over from physics, and some from even more abstract areas of maths like topology. But the overarching concept was a straight lift from thermodynamics – the concept of 'equilibrium'. In a thermodynamic system, equilibrium is reached when there is no more transfer of energy. When a hot object has cooled down to the ambient temperature, for example, it is in thermal equilibrium with its surroundings.

The great insight of the economists of the 'mathematical revolution' was that thermal equilibrium could be seen as analogous to a situation at a market where nobody wanted to trade any more. If everyone had exchanged their goods at the prevailing price, until there was no swap they could make which would leave them with a bundle of things that they preferred to the one they already owned, then in some sense this state could be considered 'optimal', because nobody would be able to improve on it. From this initial insight, plus a lot of analytical fireworks, more or less everything sprung. It was possible to provide a rigorous formal basis for the classic supply-and-demand curves that had been included in economics textbooks for the previous fifty years. The equilibrium concept and its identification with the idea of an optimal state are the basis of economics today.

How did they do it? By making one massive assumption. In physics, one electron is exactly the same as any other, as a matter of physical fact. The fundamental particles of mathematical economics, however, are commodities, preferences and people. Treating these as identical and interchangeable is a modelling assumption, not a fact; it might be right for some purposes and wrong for others. If you make that assumption, you can add them up and define a complete and consistent 'demand function' for all goods, relating quantity to price. At any one time, the combination of all the demand functions with the amount of stuff in the world and the list of who owned it would, in turn, define one single schedule of prices which was 'consistent' in the sense that nobody could do better by further trading. This was the equilibrium of the system – the lowest-energy state to which it could be presumed to tend.

Once that move has been made, the whole mathematical toolkit of physics is available to you; solving economic problems becomes a matter of cleverly defining the problem so that it can be described in these terms, and then solving a system of equations to find the equilibrium state. The mathematical revolution in economics, which was happening alongside the beginning of Beer's career, was a huge achievement. But along the way, a lot of things were lost.

In particular, the equilibrium defined in this way only describes one side of the economy – consumption and exchange of a fixed endowment of goods. It doesn't explain how that particular bunch of stuff came into existence, let alone anything about who came to own it. In order to extend the model to talk about production, more assumptions needed to be made, and they became less defensible as the research programme progressed; they ended up being neither particularly attractive as abstract mathematics, nor particularly realistic as descriptions of the messy real world of labour, capital and technology.

And the bad assumptions were an inevitable consequence of the methodology; economics was committed to atomism* because this was the step that made the physical analogy work. And one consequence of that methodological assumption is that the new science didn't have a way to describe big, complicated economic entities (like firms or governments) other than in terms of their smallest component parts (not even individual people – individual sets of preferences). The key move of Stafford Beer's cybernetics – to take a system, draw a box around it and say 'we don't know anything about the motivation or internal structure of this box, let's just look at how it reacts to its environment' – wasn't open to them any more.

Economists ended up solving this problem by denying it (a mechanism that Stafford Beer looked at in detail, concluding that ignorance is a kind of information-processing system

---

* I can feel economists getting angry here; I am an economist myself and I'm getting angry at myself. This isn't true of all economists – practically nothing is true of *all* economists. But it is descriptive of a large and important school of thought, which became dominant over the period we're talking about in many different ways. Chapter 6 treats the history and the methodological issues a bit more fairly.

of last resort). Where analysis fails, ideology steps in, and the solution the economists decided on was to fantasise an equivalent system for producers – a profit-maximising firm with an understanding of its market, a simple production process and a manageable set of decisions. It ended up going disastrously wrong, but the original intellectual sin may have been the failure to respect the integrity of the black box.

Speaking in prose

But back to Stafford Beer, whose approach seemed to pay dividends. After a while, he had developed a 'crash course' methodology that seemed to substantially raise the educational age of the troops passing through his unit. The rapid turnover of its officers also fell; the men were no longer driving them crazy.

It was around this point that Beer's superiors informed him that there was a name for what he'd been doing: 'operational research'. At the time, there was hardly more meaning to that term than 'applied mathematics, considered in a military context'. In the post-war period, there would be an attempt to apply the lessons learned from wartime operations research in civilian life, particularly in business management. The reasons why this project never really took off are complicated, but for Stafford Beer in 1947, the main significance of this discovery was that he was moved away from his soldier-student-patients and spent two years conducting 'human factors' research for the Royal Engineers.

A much greater epiphany came three years later. Beer had left the army in 1949 and started a new career as a management trainee at United Steel, living in Sheffield. He quickly made a name for himself with his operational research methods; one of his first innovations was in the field of data visualisation. At a time when plotting a chart involved ink and graph paper, he invented a method of standardising a number of different control variables (say, the temperature of a furnace, speed of a conveyor belt, tonnage of stores and hours of overtime, all divided by their average values so that they could be plotted next to each other), allowing the construction of a kind of dashboard that would allow a plant manager to quickly detect whether anything was outside its acceptable range of variation and take action accordingly. While plugging away at this sort of administrative improvement, he read the book *Cybernetics* by Norbert Wiener. The effect was electric. He immediately wrote a letter to Wiener, telling him that he appeared to have been a working cybernetician for the last five years without realising it.

Dawn of the cybernetic moment

Cybernetics is a huge subject, and it will take several chapters to discuss it in anything like the detail it deserves. But for the time being, it's important to note that Stafford Beer wasn't unusual in having read Wiener's book in 1950 – *Cybernetics* had been the standout pop science hit of 1948, and it was still making waves. This is strange in a way, because much of the book was dedicated to dense mathematics, setting out the research programme which had grown out of Wiener's war work on automation and feedback. Even mathematicians found it heavy going. Stafford Beer, with the unusual talent that had seen him teach himself university-level mathematics during one summer at the age of fourteen, might have been one of the only people in the world who really understood it. At least, Wiener certainly seemed to think so.

The part of *Cybernetics* that captured the public

imagination was described by the book's subtitle: *Control and Communication in the Animal and the Machine*. It was one of the first books to put forward the idea of thinking machines, and that was exciting enough for the book-buying population. There were spin-offs and copycats, including a self-help book called *Psycho-Cybernetics* which invited its readers to treat their subconscious mind as a controllable 'Automatic Success Mechanism'. The Psycho-Cybernetics system, originally developed by a cosmetic surgeon called Maxwell Maltz, formed the basis of a number of self-help courses based on positive affirmations, with Tony Robbins probably the best-known follower active today.

One of the book's most lasting cultural influences, though – the prefix 'cyber', which has since been attached to everything from computer crime to sex chat – is actually a mistake. If one were to follow the etymology of Wiener's coinage, it would be 'cybernet-ics', the word coming from the Ancient Greek for the person who steers a galley by giving orders to the oarsmen. The way we've divided the word up, it's as if we'd decided that there might be other kinds of 'tronics'.*

So, cybernetics was the study of the control of systems – or specifically, the study of the control of systems with enough internal complexity for there to exist an interesting problem of how to manage them. The building blocks of cybernetic arguments are framed in terms of relationships between elements which have to be viewed as black boxes. Norbert Wiener's book

---

* This pointless bit of pedantry is perhaps slightly excused by noting that the other half of the word 'cybernetics' was not given a clean death on having its 'cyber' prefix removed. L. Ron Hubbard decided on the name 'dianetics' for his 'modern science of the mental health' in 1950, and Wiener ended up having to ask the Dianetics Foundation to stop using his name in their literature.

gave a new mathematical system which aimed to provide a language for the kinds of statements you could make about these systems of relationships.

The trick was a classic mathematician's move; if you want to solve a problem, first come up with a method for solving an entire class of problems, and then apply it to the one in front of you. Wiener saw that the necessary leap of abstraction was exactly that of Beer's black boxes; if you define the elements by their *behaviour* rather than their composition, it doesn't matter whether you're talking about wheels and levers, electronic circuits or biological cells. You concentrate on how they act, how they cause changes in other elements of the system and how they are changed themselves.

One kind of interaction in particular was distinctive to the cybernetics approach. Its great influence in modern life is the concept of 'feedback' – the cyclic and recursive use of the output of one black box as the input of another, then back again, was the signature of the systems that Wiener studied. It's a mathematical engine which produces the interesting kind of complexity – unpredictable, but not random. The word has survived, but sadly it's lost this connotation of a repeated and changing process; as used on consumer websites and by human resources departments, it really just means 'communication from the decided-upon back to the deciders, of a kind which will probably be ignored'. In a recent issue of the *Wall Street Journal*, someone suggested that it should be renamed 'feedforward', so that it didn't sound so negative, which is enough to make you weep if you care about that sort of thing.

After his initial contact with Wiener, Beer's personal network grew incredibly quickly. Within a few years, he was a friend and collaborator with nearly everyone who was doing important work in the field. Not only that, but he had carved out for himself a position of pre-eminence in the application of the new science to industry. This was one case where there was no question that Beer was practising without a licence, however; if any person was qualified to describe someone as 'the father of management cybernetics' it was Norbert Wiener, and he talked about Beer in those terms. From the mid-1950s, Stafford Beer would have a twin-track career, developing his ideas and carrying out theoretical research while also applying them in business.

Almost irresistible forces, almost immovable objects

At United Steel, Beer started doing one of the things he was best at – building a team. The success of his chart dashboards had propelled him forward as a rising star, and he was promoted to set up a new operations research group in order to apply his methodology across the company. Beer would later claim that his team was generally able to deliver a 30 per cent improvement in productivity at the plants they were assigned to. It's not really possible to verify this against company accounts because he never specifies the units of measurement, but it's clear that the top management of United Steel believed in him.

The research group continued to expand, recruiting academics from disciplines as diverse as mathematics, psychology, anthropology and divinity. They had an office of their own, converted from a large Victorian dwelling in Sheffield and called 'Cybor House', after the Cybernetics and Operational Research Group.* A number of caravans were kitted out to function as mobile offices as the crack team of cyberneticians

---

* It's still called that today and houses an insurance brokerage, a wedding ring designer and several other small businesses.

were sent across the country to enhance productivity. By the time Stafford Beer resigned in 1961, it was the second largest such group in Britain, after that of the National Coal Board.

Beer's resignation was inevitable; as it became clearer that the Cybor group had ambitions beyond conducting simulation exercises and reducing stockholding levels, the management increasingly began to resist them. A book on *Practical Ironmaking* published in 1959 by the general manager of United Steel's largest plant took the opportunity to criticise the group, saying, 'In spite of the fashionable worship of such things as Operational Research, Automation, Cybernetics . . . it is believed that the iron-works will be one of the last places where the practical man will be king.' Stafford Beer returned the compliment seven years later, referring to 'the British ironmakers I knew in the 1950s' who 'sat in the ironworks, regarding the blast furnace like a woman: potent, demanding, satisfying, temperamental, unrequiting, a captivating mystery'.

Although his career tended to force him into constant conflict with these 'practical men', their reaction played an important role in his own theories and those who influenced him. The mathematical economists of the time built their models around the concept of 'equilibrium', an imagined state in which everyone had made all the trades they wanted and everyone had reached the best outcome they could find. For the cyberneticians, an analogous concept was 'homeostasis'. This literally just means 'same state', but in context it is applied to systems where feedback is present. If that feedback has the tendency to stabilise the system, oscillations will be damped and the system will tend to settle down; an example might be the gyroscopic stabilisers of a ship which react to deviations from the axis by pushing in the opposite direction. If that feedback has the opposite tendency, then oscillations and deviations will be amplified, and the system will tend to blow up; an example might be one of Jimi Hendrix's amplifiers. If the feedback is complicated and has both stabilising and destabilising elements, then things get interesting;* it might be stable most of the time but tear itself apart if exposed to the wrong kind of shock. Or it might switch from one stable state to another.

However, unlike economists' concept of equilibrium, there's no implication that a homeostatic state is optimal, just that forces exist which will try to return the system to that state when it is perturbed. Early cyberneticians were keen on building 'homeostats' from springs or electric circuits and compass needles, to show how complicated the behaviour could be, and how something that looked stable could suddenly reorganise itself in response to a seemingly minor change. In the case of organisations, it's an important way of understanding what Stafford Beer was trying to do; a large part of management cybernetics is about understanding how the relationships and links between subsystems create homeostatic forces.

Purpose

The idea of homeostasis is fundamental to one of Beer's most lasting and contentious aphorisms. Compared to the economists' concept of equilibrium, the key feature of the cybernetician's equivalent is that there's no implication that it's the result of anything outside the system. The homeostatic state is what the system returns to, though. In that sense, it's

---

* Because stabilising feedback tends to, in a sense, 'push' the system in the opposite direction to the initial shock, it's often referred to as 'negative feedback'. This isn't the same sense the person who tried to coin 'feedforward' was thinking of.

what the system 'wants'. Your assumption about how much sarcasm is intended by my punctuation here will be dictated by your philosophical views about the whole subject matter. To what extent can a system be said to want anything?

Stafford Beer's answers to this as philosophical question fill many chapters of his later books, but in his work on management cybernetics, he takes a consistent approach – there is nothing to be gained by opening the black box. If you are trying to understand a system, all you can do is observe its behaviour, and whatever you can learn from doing so is all there is to know about it.

Considered as a general principle, this doesn't seem too disturbing. The disquieting effect only emerges when you consider a particular case, where the system under analysis is made up of human beings and the forces operating on them are their own individual desires. We can all think of cases where organisations systematically deliver outcomes that are wildly at odds with their stated objectives. Stafford Beer's cybernetics tells us that in these cases, while people's opinions are important, the facts of the organisational outcomes are what we need to work with. In his most pithy formulation of the principle, he expands the black box principle to a rather more uncomfortable statement.

The purpose of a system is what it does.

This version was taken up by management cyberneticians, and often referred to by its acronym, POSIWID. Returning to the case studies of the last chapter, the purpose of academic publishing is to generate citations for academics and profits for publishing houses; the individuals involved might want to expand and communicate human knowledge, but the system behaves in ways that support its goals, not theirs. The purpose of the pet hotel at Schiphol Airport was to shred squirrels; everyone involved may have cared about animal welfare, but the system had been set up to ensure compliance with import regulation at the lowest cost possible, so that's what it did.

It's easy to mistake the POSIWID view of the world for a slightly tiresome kind of cynicism. To say that the purpose of the system is what it does isn't to make any statement about the intentions of the people working for it. The danger of confusing the properties of the system with those of its members is one of the most important reasons for not opening up a black box. Unfortunately, it's a very common confusion; very few people are able to take a step back, view their own organisation as if from outside, and realise that they are structurally producing results which are exactly the opposite of what they had intended.

The Road to Chile

This is the big paradox of management consultancy, and it probably accounts for a lot of the bad vibes surrounding that much-misunderstood profession. If a management consultant is capable of achieving anything, doing so will involve explaining to a group of people that the way in which they are organised is stopping them from achieving the goals that they are aiming for as individuals.

But as soon as you describe the problem in those terms, you'll trigger psychological self-defence mechanisms. If there weren't strong homeostatic forces within the organisation, it would either solve its own problems or fall apart entirely. But in most cases, those homeostatic forces will work against any attempt to change what the system does. It requires an extra jolt of organisational energy, or an undeniable crisis, to overcome this self-perpetuating tendency to continue doing things in the same way.

Stafford Beer spent his career trying to design systems and organisations that were capable of adaptation. He was headhunted from United Steel by a French company, which wanted to set up a European network of management consultancies. The British part of this network, SIGMA (Science in General Management), quickly developed a management culture of its own. Staff were given regular sabbaticals to take on any project they liked, as long as it had nothing to do with work. Beer hosted regular all-hands meetings called 'Sigmoots' at which anyone in the company, regardless of seniority, was encouraged to draw attention to a point they wanted to make by banging on the table with their shoe.

Throughout Beer's written work on management, this kind of meeting is always emphasised: unstructured, informal connections between staff at different levels and performing different functions. Some of the ephemera – particularly his emphasis on the importance of having adequate supplies of cigars and whisky to facilitate conversation – might seem comically dated, but the idea of ensuring that there are links across the organisation to spread information and build consensus is entirely modern.

Beer was trying to adapt hierarchical systems so they were fit for their purpose in a changing world. He saw a strict command-and-control approach as dangerously inflexible, while excessive delegation would destroy the organisation's ability to act as a coherent system. The fundamental relationship between management units in the kind of structures Stafford Beer designed was a 'resource bargain'. A unit would be allowed to operate autonomously, but only to the extent that doing so did not jeopardise the broader system, whether in terms of a financial budget, physical resources of space, managerial time and bandwidth or general goodwill.

Bargains were made with respect to all kinds of resources under the control of the corporation, and then they were expected to be kept to unless and until the situation changed, in which case they were renegotiated. It's a concept that might be most easily illustrated by thinking of a military platoon that's given an objective – to capture a bridge, say. There is a higher level of the hierarchy that is responsible for coming up with objectives, fitting them into an overall battle plan and making sure that different platoons don't get in the way of each other. But once the orders are given, capturing the bridge is the responsibility of the platoon, using the soldiers and equipment assigned to the task. The general doesn't keep on managing the attack; the next piece of communication he or she expects is either that the bargain has been kept (the bridge is captured), or that some external factor has interfered, and the resource bargain needs to be renegotiated by the survivors.

This management theory is based neither on control nor on delegation, but on *accountability* between the parts of a business. People were able to make decisions and change their mind, as long as they could justify those decisions to anyone else who was affected. And it was a system of accountability based on the flow of information; the higher functions of the system were responsible for creating the goals, ensuring their consistency with each other and with the resources available. They then communicated their sense to the operating levels and set the system in motion. And then the system repeated, with the results of the initial plans forming the information set used to revise them.

It worked at SIGMA, but there was the usual problem of

success, familiar to modern start-ups. Beer wanted to own shares in the company, but its parent company didn't want to give them up. They offered to double his salary instead, but that wasn't enough; he ended up moving to work for IPC, a large publishing company and one of his major clients. His work there was significantly ahead of its time; he even created a subsidiary company called 'International Data Highways'.* But IPC was a big corporation, at the time the largest publishing house in the world, and the homeostatic forces were strong. Beer retired in 1970 and returned to consulting, this time as an independent operator; he wrote a number of well-regarded books and took some academic posts.

In 1972, he was contacted by someone who had read those books. In its early days, SIGMA had successfully sold services in Latin America – it had a ten-person office in Chile that worked on contracts with utilities companies and the national railway. A group of engineers and managers had avidly studied the development of management cybernetics, and even formed a reading group to study Beer's books. Just as his book *Brain of the Firm* was going to press, he received a call from Fernando Flores, Technical Director of the Chilean production development agency, CORFO; the recently elected socialist government of Salvador Allende wanted help in reinventing the whole economy for the computer age. We'll catch up later with what happened there, because first we need to think about what that might mean.

---

* It wasn't just the name that was way ahead of its time for the 1960s. His idea was to use a proprietary telecom network to transmit financial data in real time to banks. Michael Bloomberg actually did this, in 1982, and his eponymous product (the Bloomberg Terminal) is on every trading desk on Wall Street, making him, at the time of writing, the seventh richest person in the world.

Purpose and power

What is the purpose of the economy? Or, to put the question in cybernetic terms, what does the economy do? It organises the production and distribution of goods. In order to do that, it has to solve an information-processing task: specific goods need to be produced, using the available resources, and then distributed to specific individuals in different quantities. To perform this function, practical decisions must be made: matching the goods produced with people's needs, bringing together resources while minimising waste. There are ethical and political decisions: who gets what? But there is also a set of decisions to be made about the decision-making process itself: how should the process of deciding what to make and who gets it, be organised?

The fact that this is specifically an information-processing problem was something of a tragedy for the science of economics. Like medical doctors, economists are sometimes faced with having to solve problems in the order that they arrive, rather than in the logical order of one discovery building on another. In this case, the 'socialist calculation debates', over the possibility of an industrial economy organised on any other basis than private ownership and markets, became urgent after the 1917 Russian Revolution. These debates turned on questions of information and computation – they addressed the issue of whether central planning or free markets were the best way to organise the productive resources of an economy. But there was no such thing as a computer in the 1920s. Things like negative entropy, dynamic programming and even the concept of a Turing machine – the fundamental logical building blocks of a rigorous theory of information – were only invented and discovered decades later.

Addressing problems out of order is a fine way to get into a confused state; blind alleys and confused concepts might exist at a foundational level, with sufficiently massive theoretical structures on top of them that it's not possible to make things good without creating an unholy amount of institutional stress. The homeostatic forces of a field of scientific enquiry are just as strong as they are anywhere else; as the economist Geoff Harcourt said in another context, to abandon a mistaken line of scholarship altogether is 'a hard thing to have to say of thirty years of intellectual fireworks let off by some of the best minds of our profession'.

The gravitational pull of cognitive dissonance had buried the socialist planning debate by the time Stafford Beer arrived in Santiago, ready to sit down with the Chilean engineers to design a new system. Once upon a time, it had been a lively and various set of questions, but by 1972 the planning debate was a dim memory in the West and a faint embarrassment in the East. Khrushchev had replaced Stalin as leader of the Soviet Union, and the planned economy was consigned to history, an interesting idea that hadn't worked.

Multiplicity of values

There was never any chance for Stalinist central planning, but more could have been learned from the way it failed. Wiener's book had also been a hit in the Soviet Union in 1948, partly because it seemed to offer the hints of a scientific basis for economic planning. This, however, was an intellectual accident. Although economists in the capitalist world had formulated their ideas on economic cybernetics before they knew what they were talking about, economists in the Communist world made a symmetrical mistake for more or less the same reasons. 'Economic cybernetics', as Soviet university economics departments rebranded themselves, was heavily a planning discipline, dedicated to solving larger systems of equations and collating more comprehensive tables of inputs and outputs to turn into optimisation problems.

Central planning has two big problems: it's planning, and it's central. The first of these problems is perhaps the most obvious: a national economy is a big and absurdly complicated thing and the idea of writing down a huge spreadsheet or computer program to make all the decisions is an impossibly difficult task. But the second is perhaps the more wicked of the two: the problem of getting the information necessary to do any planning at all, and then centralising it in one place where the decisions are made. The problem of centralisation of information has a lot of logistic aspects, of course – the Soviet Union was always running into problems because it couldn't change plans quickly enough in response to shortages and gluts. But it also has serious conceptual problems; Friedrich Hayek won a Nobel Prize and can fairly be regarded as one of the antecedents of cybernetics* for noticing that 'information' is itself a tricky concept, and that it can't necessarily be treated like any other commodity.

Some information can be written down and aggregated, just like copper stocks in warehouses across a country; each producer knows their part of the whole, and there's a straightforwardly meaningful way of adding it all up. But although it's a measurable quantity, the total number of tonnes of copper in warehouses across an economy isn't necessarily *economically* meaningful – the importance of copper depends on what

---

* Stafford Beer records in one of his essays that he had read Hayek's work in the 1950s and thought it was excellent stuff; he was profoundly shocked in the 1970s to find that Hayek was an economic advisor to General Pinochet.

you're going to do with it, and what components it has been shaped into. Doing things like adding up different grades of wire, switches and fastenings and treating the sum as a control variable was one of the big ways that the Soviet system got into trouble. One important role of the price system is to provide some basis upon which a certain grade of copper wire in the stores of an automobile manufacturer can be compared with a different grade in a builder's yard, or with the inventory of spare switches at the telephone company.

Hayek's insight was that if the price system is providing this information, it's not clear what more the planner is meant to do. None of the individual producers needs to know about the overall state of the market for copper – they just look at the prices in front of them, and decide whether to increase production, or if they should get into an industry which doesn't depend on expensive and volatile raw materials. If the central planner is content with this pattern of outputs, then there's nothing for them to do. If they want something different, then they have a huge problem, because all the information that's being carried by the market prices at that time is exactly the vital information that can't be acquired in any other way.

'At that time' is a crucial qualification here; it's what distinguished Hayek's 'Austrian' view of the economy from the 'neoclassical' one which came to dominate the science of economics in the capitalist world. The pattern of prices, for Hayek and the rest of the Austrian School, doesn't represent an equilibrium, and they weren't interested in hypothetical stable states. For them, prices were derived from transactions, and those transactions were the way in which knowledge was percolated around the economy; it's intrinsic to the system, in the Austrian view, that it's always changing because new information is always arriving.

This is a very cybernetic concept – in fact, it's so close to the eventual 'information theory' of cyberneticians such as Claude Shannon and Norbert Wiener that it's easy to see why economics didn't pay much attention when the real thing came along, ten years after the calculation debates had been settled. It's a theory that presents the economy as an information-processing system, made up of black boxes that communicate with each other by proposing transactions at prices that are either accepted or rejected. Where it seems to differ from the more developed ideas of management cybernetics is in the implicit assumption that price signals are the only kind of signals there can be.

The story of how these implicit assumptions came to be so strongly held, and how they led to Stafford Beer's tragedy in Chile, will take a bit longer to describe. But by the time I had got this far into reading up on Beer and his ideas, my mind had already been overtaken by a pressing question: if the economy is an information-processing system, is the whole thing made of artificial intelligence?

3

Aliens Among Us

'Corporations are people, my friend . . . Of course they
are. Everything corporations earn ultimately goes to the
people. Where do you think it goes? Whose pockets?
Whose pockets? People's pockets. Human beings, my
friend.'

Mitt Romney, 2011

If the economy is an information-processing system, does
that mean that every corporation is an artificial intelligence?
If people are worried about out-of-control AI taking over the
world and destroying everything, shouldn't we have been trying
to do something about them at least seventy years ago – and
probably more like two hundred?

This seems to be where you can end up if you start thinking
about cybernetics and black boxes. Consider the example of
a young person going down a YouTube rabbit hole. There's a
human intelligence taking in the content. There's also an algorithmic system at work, showing them one video after another.
You wouldn't necessarily call the algorithm an 'intelligence',
although it does seem to do things with a purpose. But then
there's a third entity in the background – the company that
owns YouTube, and the structure of cause and effect that
brought the other two things together.

There's no one person, or group of people, making the decision about what videos our hypothetical young person is being shown. In fact, the executives in charge of the parent company are sometimes horrified and distraught at the decisions made – in 2017, there was a scandal at YouTube when it was discovered that people were producing parody cartoons featuring beloved characters burning down houses or undergoing painful dentistry, and that these were being shown more often to innocent children than to the ironic adult consumers that were their intended target market.

But somewhere, at some point in time, it's been decided that 'engagement' is the purpose of the system – what it does – and somewhere else, a set of decisions have been made about what methods are going to be used to achieve that purpose.

The third thing in the room

Maybe this third entity – the corporate owner of YouTube – is the non-human intelligence, and the algorithm is just one of its component parts. Maybe another, higher-level algorithm is made out of policies, employee handbooks and corporate communications. And if that's the case, the hardware that's running this algorithm is mainly made up of people, sending emails and holding meetings, conveying information to one another but shaped by the overall structures. Are corporations people? Are they made of people?

Oddly enough, this is an idea that keeps getting independently invented by all sorts of different people. The science fiction writer Charlie Stross, for example, described corporations as 'very old, very slow AIs' in 2017. A computer science professor called Ben Kuipers wrote a paper for a conference in 2012, in which he made the case that corporations met all the criteria necessary to be called independent beings, and that as such, they were artificial, intelligent and surprisingly successful in evolutionary terms.

The basis of the idea is that 'an algorithm' is a mathematical concept – and it's meant to be neutral with respect to the system that implements it. You can test one with pencil and paper, simulate it with pebbles and buckets or hand it over to a room full of people. If a corporation has a set of rules that take inputs and produce outputs in a systematic way, then that's an algorithm. And the company could be seen as the system that implements that algorithm, in just the same way that a computer runs an artificial intelligence program.

But is this a metaphor, a literal truth or something else? After all, corporations aren't *really* people, are they? My friend Calum Chace, the futurist and transhumanist, put it this way. If a corporation is an AI, then is a family? How about a football team? Three guys sitting at a table in a pub? At some level, if you care about artificial intelligence, you need the term to have some content. What does it really mean for something to be 'artificially intelligent'?

Boxes of different kinds

At one time, it really mattered to me that computers had to be literally intelligent. At the age of around sixteen, my mind was blown by the idea of 'strong AI', that computers or robots might have thoughts, feelings and consciousness just like humans. If I had clear memories, they would no doubt be hilarious – it's not exactly difficult to come up with uncharitable and mocking explanations of why bookish, shy young men during the hormonal explosions of adolescence might suddenly get into the idea that machines might love them back. But I don't have a vivid memory; all I have is a load of old notebooks in a cardboard box full of intense philosophical essays, some of which were later inflicted on university professors and contributed to my nearly failing a term. I didn't even own a computer at the time, so they're all written out longhand. Going on about this obsession certainly cost me a girlfriend; that's a measure of how much I was into it.

A significant proportion of those notebooks were filled up with angry refutations of a philosopher called John Searle and something called 'the Chinese room argument'. By this argument, Searle aimed to prove that there was no such thing as artificial intelligence, or at least not artificial intelligence based on an algorithm.

The thought experiment is easy to summarise briefly. Imagine that we are taking advantage of the abstract nature of an algorithm, and testing an artificial intelligence program by running it by hand. Imagine that the program has been implemented not on a computer but as an employee handbook – a set of instructions for a single bureaucrat living in a room. In Searle's experiment, it's an algorithm which allows the system to communicate in Chinese, to a level which would convince a native Chinese speaker that they were conversing with another Chinese-speaking human being. Because we're not dealing with a computer program, the algorithm has to form a set of instructions to the employee, telling him how to recognise sets of Chinese characters posted through the room's letterbox, and how to find different sets of Chinese characters in his filing system to push back through the same letterbox.

We are meant to ignore lots of obvious logistical problems related to the fact that this is clearly impossible, because this is a thought experiment. Searle's point is that even if we wave away the practical objections, the conceptual one remains – the bureaucrat and his instruction book form the 'algorithm', and although he can push cards through letterboxes and pass the test, the book hasn't taught him to speak Chinese. There's nothing here, in an algorithmic system that by hypothesis acts like it has human understanding, that has any subjective experience of the understanding it simulates.

There are dozens of objections to this argument, some of which I came up with in my teenage journals. More or less everyone in philosophy of mind seemed to agree that it was wrong, although nobody could agree why. But with the perspective of thirty years to forget about how much it once mattered to me, my teenage journals suddenly fit into a pattern. Whatever else it is about, the Chinese room argument is partly about accountability and partly about respecting the integrity of the black box – the signature move of management cybernetics.

If the system is comprehensible, there's no significant problem of accountability. It may be logistically difficult to talk to the decision maker, but the focus of accountability isn't hard to identify. On the other hand, if the process which produces the output is inscrutable, then the output has to be regarded as being *from the system*. In Searle's example, if the answers given by the Chinese room are wrong or malicious, then there's no way of holding the operator inside the room responsible – he doesn't understand what's on the cards.

For some philosophers, this is the key to what it is for a system to be considered intelligent. If we adopt the 'intentional stance' towards something, we treat it as if it has mental states and attribute the things it does to its intentions. Other philosophers disagreed, of course, and said that we must ask whether the system actually has mental states, experiences and wishes. But maybe we could ignore this?* After all, whether or not there

---

* Or indeed, 'leave it to the philosophers', a phrase which more or less means 'ignore it'.

is any kind of intelligence present in a system, real or artificial, there definitely are decisions being made.

Decision-making systems and accountability sinks

So, from here on in, I will try to refer to 'decision-making systems' rather than 'artificial intelligences'. Corporations are systems, and they make decisions, so they're decision-making systems. The question is whether they're black boxes or not – whether we are able to attribute the actions of the corporation to individual human beings within it.

This is the same question that we saw in the first chapter, when we were talking about accountability sinks. There's a strong emotional pull to look inside the black box – we want to hold individuals responsible, send bankers to prison and make the chief executive of BP apologise for leaking oil into the Gulf of Mexico. But as we saw, there are organisational structures set up to prevent us from doing this.

Reading Stafford Beer and realising that I could treat decision-making systems as black boxes had begun to relax some of my previous prejudices about accountability sinks. The general flight from accountability wasn't necessarily being caused by the sneakiness of professional managers, or the psychological and legal considerations that made taking responsibility intolerable. It was more likely that managers didn't feel accountable for the actions of organisations because it didn't seem to them as if they were the actual decision makers.

Why not? Because things were incomprehensible to them too. Even moderately complex interconnected feedback systems have so many working parts that they have to be understood as a whole or not at all.

One of Stafford Beer's mentors, the psychologist and cybernetician W. Ross Ashby, put it this way:

> When there are only two parts joined, so that each affects the other, the properties of the feedback give important and useful information about the properties of the whole. But when the parts rise to even as few as four, if every one affects the other three, then twenty circuits can be traced through them; and knowing the properties of the twenty circuits does *not* give complete information about the system. Such complex systems cannot be treated as an interlaced set of more or less independent feedback circuits, but only as a whole.

What this means is that when you have to think about connections between parts, rather than just counting the individual parts of a system, the number of possibilities grows very rapidly; the potential combinations multiply, rather than just adding up. Very quickly, they multiply up to astronomically huge numbers, spelling absolute death to any hope of knowing the entire state of the system.

In fact, knowing only a few of the feedback circuits can be actively misleading, if you rely too greatly on your partial information. It is a sobering thought, for example, that despite employing some of the best and brightest* analysts in the world, the advice given by the US State Department over the last fifty years could comfortably have been outperformed by a

---

* The phrase 'the best and the brightest' is often used by people who don't know that its original context was ironic. It entered the language as the title of David Halberstam's book about the policy mistakes of the Vietnam War.

parrot that had been trained to repeat the phrase, 'Don't start a war.' The repeated failures of the State Department are not the consequence of ignorance; they are the consequence of having very good and deep – but not total – knowledge of an extremely complicated situation, in which facts outside of that information set turned out to be crucial. Knowing a great deal of detail about a subset of a system has a habit of increasing your confidence in your opinions disproportionately from their reliability.

This is also a common way for financiers to lose money. If you think of the cryptocurrency bubble, for example, it was far safer to remain ignorant than to learn enough about the stuff to be tempted to buy it. A proverb attributed to the investor Sir John Templeton holds that 'the most expensive four words in finance are, "It's different this time,"' and specialist knowledge of a new technology is the way that you convince yourself to forget about the regular black-box behaviour of the market. I'm reminded of something that a fund manager I used to sell to once told me about the global financial crisis of 2008.

'Dan,' he said, 'since this thing began, there have been two types of analysts. Some people, like yourself, have been trying to develop their understanding of an incredibly complicated system, under huge pressure, absorbing vast amounts of technical detail in a short time, and doing a fairly good job of it. Others have just been mindless bomb-throwers, trying to attract attention to themselves with ill-informed displays of competitive panic. I decided early in this crisis that I was going to listen to the *second* type of analyst, not the first – and they have turned out, systematically, to be much closer to being right.'

Not even the CEO

Here's the thing: working inside a corporation (or any large organisation) is the quickest way to realise that you have only a partial understanding of how it works. You find yourself involved with decisions, but you know that you make them on the basis of collective understanding in line with policies, with regard to the sensitivities of other divisions, based on the information provided, and so on. There are amazingly few occasions in everyday business life when someone makes a specific and important decision as a creative act. The daily grind of working life is the selection of the option that looks least obviously disastrous, according to a set of criteria laid out in a plan that was produced elsewhere.

No wonder people create accountability sinks. If your workplace is a small or medium-sized company with no conflicts of interest among its owners, you might be able to understand or predict its activities; if it's any larger or more complicated, you'd go crazy trying. For a large organisation, everyone ends up having to adopt something like the intentional stance, for the reason that Ross Ashby gave: 'Complex systems cannot be treated as an interlaced set of more or less independent feedback circuits, but only as a whole.'

This suggests that there might be a solid theoretical basis in cybernetics for the 'principle of diminishing accountability' that I thought I'd discovered, then given up on, in the last chapter. The extent to which individual accountability can exist in a system is dependent on the extent to which you treat it as a black box, which you do when something is too complex to deal with any other way.

And therefore, the observable fact that institutions tend to develop so as to reduce the amount of individual accountability is a consequence of the less controversial fact that institutions tend to get more complicated as they grow. This means that we can answer Calum Chace's question about whether there's any meaning to the idea of treating organisations as artificial intelligences. Nobody I've talked to really believes that Boeing is an intelligent being in the sense of being conscious and having independent will. But it's not *just* a metaphor; it's a simplifying assumption which is appropriate in many contexts. When Boeing does something – say, delivers an aeroplane like the 737 MAX with a fundamental flaw that causes it to crash in specific common circumstances, it's usually more sensible to say that it was Boeing that did this than to take either of the alternative approaches – making a list of all the different decisions that went into that decision, or finding some poor middle management body and pinning all the responsibility on them.

Conspiracy versus cock-up in a cybernetic framework

If you can make that mental leap, a lot of modern life becomes more comprehensible, at the cost of being a bit more bleak. Putting together two conclusions, it seems that in many circumstances, the only sensible thing to say about a lot of important events is that the actions which led to them were carried out by organisations rather than individuals. But we don't believe that organisations have interests, desires or wishes. They are black boxes; their purpose is just what they do.

That means that there's often no point in asking whether something was a conspiracy or just an unfortunate error. When an individual person causes harm, we might ask whether they did it out of a misunderstanding of the facts or of their own interests, or whether they did it because they wanted to and it served their interests. The answer to that question will be relevant to the degree of responsibility we assign – it's how we establish accountability.

An organisation does things, and it systematically does some things rather than others. But that's as far as it goes. Systems don't make mistakes – if they do something, that's their purpose. But it also works the other way round. Systems don't have inner desires, so they don't do things intentionally either. There's just a network of cause and effect. We might think they're conspiring, but they're working within structures that made the outcome inevitable. Or we might see everything as a terrible cock-up, but we don't understand that the outcome was the inevitable result of the way the system works.

That's the picture you get from reading the internal emails from Fox News about that voting machine company. There didn't appear to be any wider scheme; certainly, nobody seemed to be making a conscious decision to weigh the commercial benefits of defending their market share with a particular audience against the risks associated with broadcasting massively defamatory statements about someone with the resources to sue. All the individual decision makers were just reacting to their immediate pain points – looking at audience figures and share price movements, reading angry viewer feedback. Everyone assumed that someone else would make sure nothing too bad happened.

But nor is it possible to say that it was *just* an unfortunate error on Fox News's part. The company hadn't got into a situation where its audience was fired up with beliefs about stolen elections by accident. The local decisions to stir up a bit of rage for a quick ratings win had cumulative consequences. There was an institutional, system-wide decision made, possibly without any one person realising they were making it, that this broadcasting organisation would do what the angriest and least-informed part of its viewership wanted it to do. It was a decision that nobody made.

It's not what you'd call an attractive view of the world – most of us are just cogs in a machine, working on what's in front of us while the big picture is determined elsewhere. It's particularly unattractive if you're a CEO or head of government, and you feel like you ought to be making the decisions. But it feels comprehensible, in tune with the way things seem to keep turning out. Among other things, it might provide some sort of explanation of why it is that public inquiries into major scandals always seem to be so unsatisfactory.

Think of the investigation into the Grenfell Tower disaster, or into the Bloody Sunday massacre, or into child abuse in the Catholic Church. They go on for ever, cost huge amounts of money, and often end up producing results that fail to give the people affected the closure they were looking for. They're unsatisfactory because an inquiry is partly an attempt to investigate causation in a system too complicated to understand, and partly because it's an attempt to assign responsibility where it's hardly applicable. Sometimes, they are able to understand enough about a decision-making process to establish that one or two particular people might have enough responsibility for the outcome to be criminally liable, or at least to be worth criticising by name in the report of the inquiry. But more usually, they produce nothing more than heavily caveated lists of recommendations, after a period long enough for the organisations involved to claim that everything has changed.

I worked on the design of public inquiries in my first job, at the Bank of England, and I have to admit I completely failed to come up with any useful alternatives. If anyone wants to spend time and effort on doing a better job than I did, though, I'd advise them that a lot more money and prestige is currently available to those who apply the same problem to a different set of black boxes.

The duty and problem of explainability

The idea that corporations are frightening alien intelligences has to be treated with scepticism, but the idea that our lives are increasingly being ruled by merciless and incomprehensible decision-making systems is practically mainstream. Everyone's aware that more and more important decisions are being taken by algorithms and artificial intelligence, and it seems that unaccountable decision-making is definitely regarded as a bad thing when computers do it.

Computer-based artificial intelligence is on the cusp of change. Nearly all the actual systems being marketed under the name of 'artificial intelligence' (or 'machine learning') are not black boxes as Stafford Beer and Ross Ashby understood them. Instead, they're statistical models with a structure that's comprehensible to anyone who understands the maths: complicated but not complex in the relevant sense. If you're looking into the guts of one to find out how it works, you might end up drawing a lot of ungainly spaghetti diagrams, but you have a few big advantages.

First, you know what data it's taking in – unlike a human being or a corporation, it's not going to suddenly develop new interests or draw on new sources of information. You can have access to its complete information environment, if you are on sufficiently good terms with the person who invented it. And second, the algorithm itself has to be run on an actually existing computer, which limits how difficult it can be. The heart of a modern machine learning system is usually a big matrix where most of the entries are zeroes, and a lot of the science of 'data science' lies in the invention of techniques to exploit that fact and carry out your calculations quickly enough to be useful. This also reduces the actual complexity of the thing, usually to a level where it's possible to understand.

Your biggest advantage in trying to understand an artificial intelligence system, as of the start of this decade, is that, unless it's a very proprietary financial trading system or something financed by the US Department of Defense, it's likely that it's nothing of the sort – it's just a linear or logistic curve-fitting model, the sort that you learn about in undergraduate textbooks.

But this isn't going to be true for ever. And 'existing computers' become less limiting every year. Even the assumption that you know what the inputs are is not rock solid.

At some point in that strange tense that might be called 'Silicon Valley near future' – the way that one has to talk about things which have been theoretically possible for a while, are actually being done in a small set of applications right now and will be ubiquitous within a decade – there will be lots of computerised decision-making systems that cross the black box threshold of comprehensibility, where, in the words of Ross Ashby, they 'cannot be treated as an interlaced set of more or less independent feedback circuits, but only as a whole'. And at that point, looking through the computer code line by line isn't going to help anyone understand why they're getting the answers they are.

Looking ahead to this time, the people who write these computer programs – and the people who know enough about the subject to be scared of it – have started to believe that there is an urgent problem of 'explainability'; a sufficiently complicated computerised system will soon come up with decisions where it is not obvious why you got *that* answer given *that* input data. The artificial intelligence guys also talk about the 'responsibility gap', which refers to the problem of assigning liability when, say, a self-driving car kills a pedestrian or when a scoring system turns down every single loan application from a Black neighbourhood.

I spoke to a statistician about this. He pointed out to me that explainability is itself a somewhat biased concept, because it's rarely analysed as a general problem of explaining which inputs are most likely to change the output. To put it bluntly, nobody is particularly interested in studying the explainability of nice decisions – when people are working on the subject, they're usually doing so in order to explain why someone has been turned down for health insurance.

This means that one big reason why research around explainability is well funded is that it has important applications even to the relatively simple algorithms available today. For example, we often need to prove that a computer isn't racist. When statistical learning systems are used, they are trained on past data, which often relates to past decisions made by human beings. And if the system learns to replicate human decision makers, it will often accidentally learn their biases – Amazon, for example, ended up having to get rid of an early algorithmic decision-making system that they were using for employee recruitment because it was trying to match the data set of successful employees on which it was trained, and consequently systematically favouring men over women.

Ethics of AI is business ethics

There's a disconnect here, though. Amazon identified this feature of their algorithm because it wasn't a black box – it was a simple enough piece of software for them to track the connections and prove mathematically how it was making its decisions. The 'explainability' project is meant to ensure that this property will be preserved as the algorithms get more complicated. And looking at it, you can see two separate work streams. There's a technical programme of devising techniques to get 'good enough' explanations when opening the black box and looking at the computer code isn't going to work. But there will also be a substantial pool of career opportunities available to humanities graduates in the 'ethics of Artificial Intelligence', devoted to answering abstract questions about what sort of explanation we might accept. There's even work for lawyers, because in some contexts explainability is a legal right; in the European Union, for example, consumers of financial services have the right to know that no decisions are taken about them by a computer without human intercession.

So explainability in artificial intelligence is a big and important problem, both technically and ethically. But hold on. When a decision is made by a corporation without any computerised process, what kind of rights to an explanation do you have? Very few.

You might, if you were lucky, get a copy of the relevant corporate policy. It would be difficult to get internal emails and notes of meetings, but in the best case, a very well-resourced official inquiry with subpoena powers would be able to get them (if they haven't been destroyed). But even then, a lot of the factors influencing decisions are tacit and implicit – the practices that translate the dry corporate handbooks into real life, the shared understandings of what constitutes important information, and so on. You've got no chance of getting access to those. That explains why a middle-ranking technical pilot called Mark Forkner was the only person who faced criminal charges over the Boeing 737 MAX, as if the inaccuracies in his reports to the aviation authorities were the sole cause of the deadly crashes. He was acquitted, by the way, and no prosecutor has to date taken on the notoriously difficult task of proving criminal intent on the part of a corporation. So, it looks like the corporate accountability sink has performed its function.

This means that the problems of the 'ethics of Artificial Intelligence' are already with us. Computer algorithms are now close to becoming black box systems, and we instinctively think that's problematic – but organisations of all kinds have been working at that level of complexity for centuries.

With respect to the organisational black boxes called 'governments', a lot of thought has been devoted to the nature of legitimacy, the extent to which decisions can be challenged and the amount of transparency we should expect. With respect to the organisational black boxes called 'corporations', much less so.

Paperclip maximisers

Explainability is far from the most important way in which the problems with hypothetical artificial intelligences overlap with the problems with real non-human decision-making systems. An even more fundamental issue that if a decision-making system isn't a human being, you can't assume that its system of priorities is going to be comprehensible or justifiable. In particular, consider a system that was simple when it was created, but then grew more complicated.

A 'paperclip maximiser' is one of those thought experiments that sounds like a bit of a joke. But then you notice that people who have made millions of dollars out of knowing a lot about artificial intelligence are taking it seriously as a practical possibility.

The concept was invented by the philosopher Nick Bostrom to illustrate the difficulty of controlling something much more intelligent than you. Imagine that you wanted to make an artificially intelligent paperclip-manufacturing machine. And then presume that, in order to make stationery more efficiently, you gave the AI in your paperclip machine the ability to modify itself and improve its intelligence. The 'paperclip maximiser problem' suggests that by doing this carelessly, you might accidentally destroy all human life in the universe.

The thought experiment might play out as follows. Your machine can improve its intelligence, and it has the desire to do so; the more intelligent it gets, the better it gets at making paperclips. But every improvement in its artificial brain increases its ability to make further improvements, so it will get cleverer at an accelerating rate. By the mathematics of exponential growth, a point will quickly arrive when the paperclip machine is not only cleverer than its inventor, but cleverer by so many orders of magnitude that the difference is similar to that between a human and a fly.

This is widely regarded in the AI community as something that's likely to happen sooner rather than later. The fact that it currently seems a distant possibility is only weak evidence that it might not happen – processes of exponential growth always look slow and innocuous until they look terrifying and overwhelming. The idea of 'the Singularity' – the event at which artificial intelligence becomes the dominant intelligence in the universe – has been around for decades. It's almost as old as the theory of artificial intelligence itself, which dates back to the 1940s.

If you're going to create something that will become the dominant form of intelligent life in the universe, of course, you should have an idea what its priorities are. Because if its priorities are simply to make as many paperclips as possible, then that's what it's going to do. If the intelligence behind the paperclip factory is many orders of magnitude greater than any human intelligence, it will easily think of a way to allocate more resources than anyone had previously thought to paperclip production. The thought experiment tends to end up with all life on earth extinguished and our bodies disassembled molecule by molecule to provide raw material for paperclips.*

The analogy with global warming is so clear that I needn't insult your intelligence by explaining it. Most of the people who have written about the similarities between corporations and artificial intelligences seem to have done so at least partly on the basis of this property – the fact that corporations prioritise their financial results, and appear to be unable to change course even when faced with the imminent extinction of human life.

It seems that just like human intelligences, artificial intelligences can have pathologies. They get fixations and obsessions, and they make decisions that cause harm without necessarily understanding why. This is not just true of 'artificial intelligences' in the sense of computer programs, it's a problem that could potentially happen to any sort of decision-making system that's complicated enough to need to be treated as a black box.

In any case, 'explainability' isn't an objective fact about a situation; to state that there is a problem of explainability is to state that a decision-making system is sufficiently complicated that it needs to be treated as a black box. And this, in turn,

---

* In particularly Grand Guignol versions, we get taken apart for raw material for interplanetary spaceships intended to scour the universe for more paperclip production resources and all alien life gets destroyed too (except, possibly, if somewhere out there is a race of sentient paperclips).

means that this is still all about accountability. The decision to treat something as a black box is often not politically neutral.

The Algorithm and the system

Because of the need to cancel exams in 2020 as a result of the Covid-19 pandemic, education ministries around the world required an alternative way of awarding grades to high-school students. The way in which the Department for Education in Westminster went about this provided weeks of conversation for the English middle class. 'The Algorithm', as it became known, is a failure of automated decision-making that is likely to end up in future computing textbooks. It combined serious conceptual mistakes with an underlying model that couldn't possibly perform the task, in the service of an overall goal that didn't make sense.

Arguably the fundamental problem with the exercise was the decision to start it at all. The English system at the time was dependent on examinations – for most subjects, teacher assessment or coursework played a minimal role. If your grading system is based on exams but you can't have them, it might have made sense to consider at an earlier stage that grades might not be possible – but they were important inputs for another part of the system.

Exam grades were used to allocate undergraduates to courses, and their tuition fees were essential to the universities. So there needed to be grades, and they needed to follow roughly the same distribution as in previous years, so universities could be reasonably confident that the number of students they were asked to take wouldn't be too far away from their planning assumptions.

This wasn't the only reason why the staff of the Department

for Education were keen to ensure that the overall distribution was unchanged. The Algorithm was a classic example of a decision nobody made in a similar way that the 3 per cent fiscal deficit target emerged from a meeting in the civil servant Guy Abeille's office to become the economic law of Euroland. The department asked its research staff to come up with something. But they didn't make it clear that whatever they came up with would directly determine the final outcome, rather than being one input into an overall process that would trade off different priorities against each other.

The research staff of education policy institutions care a lot more about 'grade inflation' than almost anyone else. For most purposes, your high-school grades are insignificant a few years after you join the workforce. They allow universities to differentiate between applicants in any one year, and employers might use them for the same purpose. But it's quite rare for anyone to have to assess the merits of two candidates who did their exams in different years, when they don't have lots of other information about them that's much more important than their grade in A level Chemistry.

Almost without anyone noticing, the algorithm that was designed was a sort of 'paperclip minimiser'. In the technical documentation, it seemed that the main design criterion had been to ensure that the overall distribution of grades was as close as possible to the curve describing the allocation from previous years. This would minimise the degree of grade inflation and in the minds of the designers, it was 'fair on average'. But of course, you can't be fair to an average; fairness or unfairness happens to individuals.

The Algorithm took the assessments given by schoolteachers as inputs and then increased or decreased them, based on an equation that was meant to estimate the extent to which the school might have overestimated (or, more rarely, under-estimated) the grade that pupil would have attained had the exams taken place. This equation, in turn, was based on the proportion of each grade that had been earned at that school in previous years.

If anyone had thought about this, they might have real-ised that a system that penalises bright young people for going to school in underprivileged areas was always going to raise serious concerns. But they didn't. This appears to have been at least partly because of an offsetting error. Where there wasn't enough historical information to define the parameters of the equation, the Algorithm ignored it and used the teachers' unadjusted grades. So, in the name of fairness, the DfE ended up creating a system that rewarded small exam classes (more common in private schools) and penalised large ones with an undistinguished track record. But it performed well in terms of the closeness of fit to past years' curves; it was fair to the average.

On the day the adjusted grades were announced, all hell broke loose, demonstrating my friend the statistician's asser-tion that explainability only becomes a problem when you're giving bad news. There was a lot of bad news, particularly for inner-city kids with dreams of going to university. When the DfE published the details of the Algorithm, in the hope that everyone would understand why they had done it that way, things got considerably worse. Not only were technical errors quickly spotted, but people began to understand that the overall approach was aimed at minimising grade inflation – and that the decision to sacrifice numerous individual lives to this goal was built into the system.

At this point, the Royal Statistical Society announced that they had offered help at an early stage in designing things, but that the DfE had demanded such draconian non-disclosure agreements – prohibiting professors of education from saying anything critical for five years – that none of the country's qualified statisticians felt able to get involved. The department panicked and scrapped the Algorithm, returning to teacher-assigned grades and tolerating grade inflation. And people started asking questions of explainability and accountability at a higher level – how could a government department screw something up this badly?

No satisfactory answer to this question was ever given. Everyone involved seemed to have been performing the task assigned to them to the best of their abilities. Nobody ever gave the order to optimise the system to fit to the curve – but nobody ever suggested that there might be other possible criteria of fairness. The minister responsible defended his senior staff before firing them. He was later knighted.

How systems go bad

We might have got more answers if there had been a full commission of inquiry, but as I said before, I doubt it. If you read a lot of British political memoirs, it's quite a common theme that people who have had detailed involvement with the Department for Education are filled with admiration for the talented and public-spirited people who work there, but absolutely despair of the department itself. It's an organisation with a strong culture and ethos, and it's extremely resistant to change and external pressure. Every day, its civil servants are asked to perform an extremely difficult and important task, usually with wholly inadequate resources, and despite trying their hardest, the results are usually bad.

The people in the system are good, but the system itself is

not. This happens a lot. If we are stuck with the metaphor of organisations as artificial intelligences – and it looks like we might be – then we might as well continue to use it. If non-human decision-making systems exist, can they have mental health problems? And can they be treated?

INTERMISSION

Computing Ponds and Rabbit Holes

Stafford Beer had a 'computing pond', which he described in a lecture once:

> Many experiments were made . . . Iron filings were included with dead leaves in the tank of *Daphnia*, which ingested sufficient of the former to respond to a magnetic field. Attempts were made to feed inputs to the colony . . . However, there were many experimental problems. The most serious of these was the collapse of any incipient organization – apparently due to the steadily increasing suspension of tiny permanent magnets in the water.

It's not completely clear from the context whether it was a joke, an exercise in meditation or just a pet pond, but it seems very likely that his experiments were at least semi-serious.

The idea was to manipulate the pond so that the initial conditions in the algae and insect larvae it contained could be made to correspond, in some way that was never quite specified, to a computing problem. The hope was then that the evolutionary process of the pond life could be made, in some even less specified way, to develop so as to represent the solution.

Beer also worked with fellow cybernetician Gordon Pask to try to teach a petri dish to grow an ear. It was a feedback circuit which sent electric currents that would either stabilise or destroy the crystalline structures which grew in a ferrous sulphate solution, until the dish was 'tuned' and able to distinguish between different frequencies.

The cyberneticians were always doing things like this. Building electronic 'turtles' which either drove themselves towards light sources or fled away from them. Creating circuits which stepped through different voltages at random until a set of four compass needles was held in floating equilibrium.

Stafford Beer even devised a machine that would teach children to solve simultaneous equations by turning on a green or red light and assisting with their trial and error. It was a quest for what Beer called 'fabric': something which had the properties of a computer, but which didn't depend on someone writing down an algorithm ahead of time.

The dream of AI research is still of the prospect of 'getting out more than you put in'. In fact, 'explainability', for all that it's seen as an important ethical goal, is practically a contradiction in terms when applied to a system that you also want to see as intelligent.

A computer might be able to multiply two seven-figure numbers and find the cube root of their product, but this isn't really something surprising; you know that it's just iterated some very simple arithmetic operations.

Even when the computer is finding patterns in data which all human analysts have previously missed, much of the time it's doing so by applying a method which could in principle have been carried out with pencil and paper if there was unlimited time and resources available to do so.

In other words, there's no real magic to an 'artificial intelligence' if it is explainable in this sense. The process of explanation is intrinsically demystifying – it's the means by which you explain how the conjurer put the rabbit into the hat and then pulled it out.

But the possibility of something like real magic is a huge part of the attraction of AI. The dream has always been to create a machine that can come up with original thoughts, have ideas that its creators *couldn't* predict and *can't* explain. If you have a system that doesn't look like something with a comprehensible logical structure, and it starts producing logical decisions and answers, there's much more of a sense of wonder than you get from a calculating machine that produces calculations.

That's clearly the case when you're talking about a pond that can solve differential equations or a dish of ferrous sulphate learning to identify frequencies.

But even with digital computers, there are algorithms and ways of thinking about programming which make it much more difficult to see where the rabbit went into the hat than others. They are usually not the most efficient or the quickest ways to get problems solved, but because of that spooky, fabric-like nature, they're the ways which hold out the fascinating possibility of doing things which can't be done at all with conventional methods.

That's why some people have always got more excited about doing computer science that way. And this has been how it is since before the invention of anything we'd recognise today as a computer.

PART TWO

PATHOLOGIES OF THE SYSTEM

4

How to Psychoanalyse a
Non-human Intelligence

I have an uneasy feeling, for instance, that if the
computer had been around at the time of Copernicus,
nobody would have ever bothered with him, because the
computers could have handled the Ptolemaic epicycles
with perfect ease.

Kenneth Boulding, 'The Economics of Knowledge and the
Knowledge of Economics', 1966

One of the reasons why so many of the cyberneticians were
hospitable to Stafford Beer might have been that he was doing
the kind of cybernetics that they'd been hoping for. Because
even quite early in the project, things had gone somewhat off
the rails. Cybernetics had been conceived as an interdisciplinary project, with a very large input from the medical and social
sciences. But it got unbalanced really quickly. People were just
too good at inventing transistors.

Tremors and ataxia in automated gunsights

Norbert Wiener had first thought about control systems in
the context of his war work. One of his projects at the US

government's Office of Scientific Research and Development was to invent an automated gun sight that could predict the movement of an enemy aeroplane and 'aim off' sufficiently to compensate for its likely movement while the bullet was in flight. Simply extrapolating along a straight line in the direction that the plane was heading wasn't working, as pilots would be taking evasive action. But Wiener and his team realised that the planes' movement couldn't be random. There were limits on the sharpness of turns that could be made, because of the need to maintain structural integrity of the airframe and continued consciousness of the pilot.

The set of manoeuvres that pilots used was an even smaller subset of the theoretically possible stunts. Flying an air raid was itself a hugely stressful task, leaving only a small amount of mental bandwidth for evading ground fire. So, pilots tended to make spur-of-the-moment choices from a small number of techniques; in principle, a manoeuvre might be identified from its earliest stages and the rest of the trajectory predicted. All that remained was the task of fitting a mathematical curve to the trajectory and transferring this information to the servomotors of the gun turret.

This turned into a problem of feedback; the gunsight moved, which meant that its input changed, which meant that it moved again. Quickly, Wiener found that the system had a tendency to develop two kinds of problems. Either the feedback was too weak, leaving the gun trailing behind its target, or it was too strong, and the gun oscillated around the target, overcorrecting its errors in one direction and then the other.

A friend who was a medical doctor encouraged Wiener to interpret these problems as equivalent to two symptoms of injury to the cerebellum – ataxia, and the 'purpose tremor'. This, arguably, was the moment at which the projects of control engineering, feedback and communication theory became distinctively 'cybernetics'. The idea had been born that a common quantity was preserved between the radar, the computing machine, the servomotors and the human gunner.

All the wonderful things

This quantity was, of course, 'information' – the subject of the socialist planning debate, but in a context where it might be quantified and turned into the subject of rigorous theory. And for many other scientists, partly because war work had thrown them into the company of people in other fields, it seemed like it might be the key to another scientific revolution. After all, as well as gunsights and commodity markets, information was the stuff of the human brain. It was what people wrote down in books, what organised individuals into societies, the stuff of thought itself. By 1945, dozens of people – telephone engineers, brain doctors, ecologists, anthropologists, philosophers, physicists – had bottom desk drawers bulging with ideas that had come to them when they were working on breaking codes, treating shell shock, designing radar consoles or, in Stafford Beer's case, mapping relations between tribal factions. They wanted to keep talking about them.

And so they did. In Britain, there was the Ratio Club, whose members included Alan Turing, while in America the Macy conferences* invited Wiener and John von Neumann, as well as the anthropologists Geoffrey Bateson and Margaret Mead. People

---

* The Josiah Macy Foundation sponsored conferences for scholars to present their research in progress in an interdisciplinary setting. These conferences covered a number of topics loosely related to the brain sciences, but it was the cybernetics series that really got everyone excited.

made light-seeking robots called 'tortoises' and observed that they could get into pathological states that resembled 'depression' or 'compulsive disorder'. They built mice that could find their way through mazes and 'homeostats' that could find their way back to a balanced state when their equilibrium was disturbed. And they drank, smoked cigars and argued. Over dinners and in seminars, it was a rare moment when people were both making rapid progress in mathematics and trying to think about the practical implications. Wiener's *Cybernetics* gives a flavour of how things were in 1948; the first chapter, 'Newtonian and Bergsonian Time', consists of a philosophical speculation on the nature of consciousness; the second is called 'Groups and Statistical Mechanics' and has several pages covered with nothing but equations. Unfortunately, the project failed.

The sad demise of the Cube-turning Society

This might be considered an unreasonable assessment on my part. After all, the members of the Ratio Club and the attendees of the Macy conferences between them invented information theory, the architecture of the microprocessor, digital computation and a large proportion of everything that makes up the modern world. Without the intellectual explosion in computing and information technology that took place at the end of the Second World War, life today would be unimaginably different.

But very little of their progress was in the field of cybernetics. What happened was that in trying to solve the problems of 'control and communication in the animal and the machine', people had ideas that were applicable to other problems. The solutions to these problems helped them design fantastic new machines, things which had commercial applications beyond the demonstration of simple ideas about control and representation. And within a decade or so – basically the period between the first edition of Wiener's *Cybernetics* in 1948 and the second in 1961 – most of the effort had been diverted towards the new and useful technological projects of computation and telecommunications.

What happened? Here's an analogy. Imagine a club of people who are interested in turning cubes 'the right way up'. In the early days of the cube-turning movement, they start out debating what it means for a cube to be the right way up and how 'way-upness' can be manipulated. But they are clever people, and breakthroughs come quickly; pretty soon they get the idea of colouring the cubes' faces so that they can talk about 'red side up', 'blue side down' and so on. Within a few years, they have established fundamental theorems about cube orientation, and the society announces a grand symposium, where the leading cube manipulators of the day are to discuss the new science of turning cubes.

Within a year of the symposium, the club is more or less defunct. It turned out that there were two factions within the Cube-turning Society, and that they were interested in different problems. One group was interested in making long lines of cubes, all the same way up. The other group wanted to solve Rubik's Cube puzzles.

While the question of what constituted the orientation of a cube was live, the two groups could work together. Once that question had been resolved, their underlying difference became acute, because the two problems are not the same. The group that wanted to make long lines of cubes could make rapid progress, because orienting a hundred cubes red-side-up is subject to a simple algorithm,* applied one hundred times.

---

* Check if the cube you are holding has any red faces – if it doesn't,

The Rubik's Cube group, however, would still be faced with a difficult problem. Colouring the faces of their puzzle would have given them a concept of what a solution might mean. But it doesn't instantly solve their problem – in fact, it could be argued that it raises a few new questions for them, as they would realise that their original puzzle isn't actually a stack of twenty-seven cubes but only looks like one. They now have to grapple with the question of whether their original vision of the puzzle as a cube-turning problem was valid. They certainly can't assume that the strides being made by their former clubmates in developing more efficient ways to arrange lines of red-side-up cubes will be helpful to them.

Philosophers and telephones

This, more or less, was what happened to the science of cybernetics. Some people were interested in information because they wanted to transmit messages from one location to another, or to process strings of bits representing numbers in series of sequential operations. Others were interested in systems with multiple inputs and outputs and complex connections between them – systems like brains (and, when Stafford Beer came along, organisations).

Very quickly, the physicists and engineers realised that the concept of information was related to things that had already been established in the statistical-mechanical interpretation of thermodynamics. In particular, they saw that information was a *difference* between states – a system in perfect thermal

---

throw it away – if it does, select one of the red faces at random, place the cube at the end of the line with this face upward, then pick up another cube.

equilibrium had, in some sense, lost its information. Information was therefore related to the concept of entropy, the extent to which energy becomes disordered and unusable as hot things cool down, cold things warm up and the universe tends towards inevitable stasis.

This leap of understanding allowed a great deal of the existing mathematical toolkit of physics to be brought to bear on the problem. A number of key concepts were soon developed, relating to the capacity of a noisy channel to carry a message and methods for error correction. 'Information theory' had been born. There were actually two versions of it. In Wiener's theorem, information was represented by entropy with a negative sign in front of it, and was interpreted as the amount of 'order' in a sequence. Claude Shannon of Bell Labs derived the same theorem independently, but in his version there was no negative sign, and information was taken as representing the extent to which a sequence was *un*predictable – whether it constituted 'news'.

Shannon's version of information theory is the one that more people remember these days; it was more suited to communications and computing, because (among other things) it measures the extent to which a message can be digitally 'compressed' to take up less memory space by getting rid of redundancy. Wiener was reasonably generous in allowing the credit for discovery, but stuck with his own formulation and regarded it as philosophically significant.

But the rapid progress in information theory was mainly related to operations that could be carried out bit by bit, one at a time. Progress on complex systems – the initial goal of the cyberneticians – was much slower.

The trouble with complex systems is that combinations of things tend to multiply together rather than adding up, so the number of possible states gets out of control very quickly. Even a Rubik's Cube has more than 43 quintillion possible states; clearly a brain or an organisation has far more. So one side of the initial cybernetic project was always going to develop far more quickly than the other. Pretty soon, everyone who could find a way to convert their problem into one where the bits of information could be addressed sequentially rather than in combination had done so. And the new science of communication and control was left in the dust, subject to the unfair perception that it hadn't made much progress.

The philosophers and neurologists were not just hampered by the mathematics of combinations; there was also an element of 'decisions nobody made' to the structure of cybernetic research after the war. As we have noted about economics and medicine, some sciences are cursed with the need to answer questions as they arise, rather than in a logical order. For the early cyberneticians – and therefore, for the majority of computer scientists to the present day – crucial, foundational assumptions were built into the discipline in the 1940s, largely because of the war. Wiener came to regret having decided early on that if control systems were to be used for war work, they would need to have a number of practical properties.

They couldn't depend on accurate measurement, because of the conditions they would be operating under. Information would need to be represented and calculations made digitally (like an abacus) and not in analogue form (like a slide rule). They would probably need to be implemented in electronic rather than mechanical form; electronic switches could be operated more quickly and would not wear out. This in turn meant that numbers would need to be represented as binary digits (later known as 'bits'). And since every occasion on which information was input by a human being or output for human understanding would slow the process down and introduce opportunities for errors, as much of the process as possible had to be put on to the machine.

These decisions meant that research into other kinds of computing would move more slowly, with the pace of improvement in digital electronic computing sped along by the progress of the electronics industry in making cheaper and smaller switches to represent the binary codes. It also meant that the research programme would concentrate on automating what could be automated rather than redesigning whole processes; calculating machines would be dropped into existing organisational structures as tools to speed up tasks that were already being carried out.

It is hard to say that these were the wrong decisions – transistor miniaturisation progressed faster than anyone expected, as did the introduction of computers into industry and government. But many people who had been involved in cybernetics seemed to think of it as a missed opportunity; Stafford Beer used to say that the way in which computers were used in the 1970s was as if companies had recruited the greatest geniuses of humanity, before setting them to work memorising the phone book to save a few seconds turning pages.

The brain doctors

Although they had been left behind, the true believers of cybernetics struggled on. The progress made in the theory of information didn't lead to immediate results, but it did give them a language to talk about the problems they wanted to address.

These were people like Ross Ashby and Grey Walter.* Both

---

* William Ross Ashby and William Grey Walter, to be precise. For

of them worked at a hospital on the outskirts of Bristol called the Burden Neurological Institute, and cybernetics was among the avenues they went down as they attempted to understand what was going wrong inside their patients' heads. Walter was an inventor of brain-scanning machines, while Ashby began as a biochemist investigating the enzymes released as a result of electroconvulsive therapy.* They were two of the first makers of machines to demonstrate how complicated behaviours – the tortoises who searched for light (and who became 'neurotic' if their own operating light was reflected in a mirror), the homeostat which cycled through combinations of settings to reach a balance – could arise from comparatively simple sets of connections.

The implication was that if these wonderful toys could be put together with simple hand-soldered wiring looms, surely a neural network with connections on the scale of the human brain could generate the complexity of the mind. Of all the cyberneticians, the British neurologists were most committed to the possibility of artificial intelligence (Ashby even wrote a book called *Design for a Brain*), which meant walking a tight-rope between science and metaphor. Trying to build things meant that everything had to be kept specific, but the theory of

---

some reason, cyberneticians were keen on dropping their first given names.

* The link between their medical practice and their cybernetic theories is frustratingly vague. On the one hand, it's clear from diaries and published writings that they regarded cybernetics as a possible avenue to a rigorous science of the mind. On the other hand, the sociologist of science Andrew Pickering has sorted through nearly all of Ashby's papers looking for a sign that he had a cybernetic interpretation of electroconvulsive therapy, and reluctantly concluded that the evidence just isn't there.

information couldn't straightforwardly be applied to massively connected systems. You needed to find a way of describing things that was both rigorous and representative of reality. As economists will tell you, this isn't an easy thing to do.

While the American tradition in cybernetics tended to be based on Shannon's version of information theory as a measure of the predictability and redundancy of a sequence of symbols to be communicated over a noisy channel, their British counterparts were more committed to Wiener's vision of information as the difference between order and chaos. Partly to emphasise this distinction, they started talking about something they called 'variety' – a quantity like information, but in contexts of large systems and where you're talking about control rather than transmission. A rough definition would be that the variety of a system is the number of states that it can be in; this rough definition also has the useful feature that it immediately raises the question: 'What do you mean, "the number of states that it can be in"? That sounds really hand-wavy and imprecise.'

Variety isn't an intrinsic property of anything; it's a property of the way that you're choosing to describe it. And that's the big step that the British cyberneticians had to take: rather than working with complete information, their work starts by saying how they're going to consider the system – which black boxes they're going to divide it up into, and what properties and states of those boxes will be included in their model.

Once you've made that initial step, you can use the same mathematical toolkit to talk about variety and control as telecoms engineers use to talk about information and communication. If you think about it, the capacity of a communication channel to transmit a message is clearly going to bear some sort of relationship to the capacity of a control system to respond to disturbances. This identification of information and control as twin quantities is at the base of what Stafford Beer called the cybernetic equivalent of Newton's laws of motion – the 'principle of requisite variety', coined by W. Ross Ashby.

This law states that anything which aims to be a 'regulator' of a system needs to have at least as much variety as that system. In order to understand what that means, we need to get a short way into the theory of cybernetics. But for the meantime, for a sort of high-level metaphorical understanding,* think about how you steer various kinds of vehicles. A train can really only go forwards and backwards, so it only needs a single handle. A car can make turns, so its control system requires a steering wheel to represent the added dimension. And an aeroplane needs a joystick rather than a steering wheel, because it can make two kinds of turns. Let's see where the cyberneticians took this idea.

The variety club of Great Britain

Let's recap. The way that you deal with massively complex systems, according to the cyberneticians, is to split them up into black boxes with a manageable set of inputs and outputs. The *combinations* of those inputs and outputs will still be unmanageable, but that might not matter; although a Rubik's Cube has 43 quintillion possible states, you can regard it as having six faces that pass coloured squares to one another in predictable ways when turned, and solve it like that. The important point is to get the description right in terms of black boxes, and to try to understand the relations between them.

---

* Which, to be honest, is often all you really need in these contexts. The whole damn *thing* in cybernetics is made of metaphors and high-level analogies. This isn't computer science, where you can point to the states, bits and circuits on a piece of silicon.

That's the art and science of cybernetics. The art is in describing the system – that's where the metaphors and analogies come in, where you have to decide whether the national wheat production system is more like a brain or a stopwatch. But once you have decided upon a set of black boxes, things become more scientific. Everything is driven by Ashby's law of requisite variety: a given system has the potential to achieve stability only if every source of variability from the environment is matched by an equal or greater source of variety in the regulatory system.

This concept of 'matching' is key when you move from toy systems to real ones – it's obviously ridiculous to try to calculate how many bits of information an individual adult might be taking in at any given moment, but we all know what someone looks like when they're overloaded. In many contexts, it's easier to check whether the variety of the system is greater or less than that of its inputs than to conceptualise what the actual number of bits might be.

This is the first and most important principle of management cybernetics. Stafford Beer occasionally referred to what he did as 'variety engineering'. By this, he meant the process of checking that the law of requisite variety was being respected at all points in the system where one thing was meant to regulate something else. If a manager or management team doesn't have information-handling capacity at least as great as the complexity of the thing they're in charge of, control is not possible and eventually, the system will become unregulated.

Like the laws of motion, things become complicated when you start trying to apply the principle of sufficient variety to any actual system. But the early cyberneticians were quick to discover principles of variety engineering that could indicate whether a system was going to fall apart. The first trick is to start giving names to different types of black boxes, and to the connections between them. Let's consider a tiny, simplified example first – how would we treat, in black box form, a simple system consisting of a cage with a squirrel in it?

The squirrel survival facility

For some purposes, we might want to treat the cage and the squirrel as separate black boxes, but we could also say that this is one black box, with a set of outputs consisting of measurements relating to the squirrel (its weight, body temperature, heart rate) and to the cage (the ambient temperature, light or dark, and so on). We can then define a set of ranges for the output variables corresponding to 'this squirrel is capable of surviving'.* This might be a simple list of acceptable ranges of values for each measurement, or it might be complicated if some of the variables interact – for example, if the squirrel needs it to be light some of the time and dark some of the time.

If we leave the system on its own, the measurements will mainly depend on random disturbances, and on the passage of time. Also, if we leave the system unattended, one or other of the output variables will soon drift from the survival set (meaning that the squirrel will die). Our model of the overall system will thus need another black box – the 'regulator' – representing actions taken to keep the squirrel alive.

---

* To be rigorous, I should say 'the squirrel is capable of surviving until the next time readings are taken'. Matching the frequency of observations to the frequency with which things can happen to the state of the system is an important part of cybernetics; constantly referring to it would make the explanation too much to take in, but it's worth remembering and it will be given its due importance when we start talking about practical implementations.

The regulatory system might consist of many components or just one – cybernetically, it's just a black box with the output of the squirrel system among its inputs, and the inputs of the cage system among its outputs. As long as we are only thinking about the system in its role as providing regulation for the cage, there is nothing gained from opening the black box. (If we were thinking about it in the context of a wider system – say, as part of a university research lab which needed to think about budgets, heating bills, personnel policies and so on – we might be interested in the nature of the components, considered as part of a different system.)

The system is 'stable' or 'regulated' if the regulator is able to keep the outputs within their 'survivable for the squirrel' ranges. We judge its ability to do this by comparing the variety of the potential disturbances with the variety of the regulator.* Some of these comparisons are easy – the lighting environment has two states, 'light' and 'dark', and the squirrel needs roughly equal amounts of both, so the regulator needs a light source with one bit of information (a switch, on and off). When the variety of the regulator matches or exceeds that of

---

* It ought to be obvious that this is going to mean that 'stability' is not an intrinsic or objective property of anything physical in the real world – it's a property of the way in which you've decided to analyse the system, what you have decided are the crucial variables and their survival sets, and so on. Also, there are always going to be some kinds of shocks that overwhelm the system. But this obscures a much more important point about stability: there is a big difference between systems that are only stable for a known range of potential shocks, and those which can adapt to at least some shocks outside the range anticipated by their designer. Ross Ashby called this second kind of system 'ultrastable', which turned into a kind of parent concept of what Stafford Beer called 'viability', the potential to persist as an independent entity.

the environment, it's possible to regulate the system – in our squirrel cage, whether it's light or dark is determined entirely by the desired value of the output, with no influence from the environment at all.

Representing the ambient temperature in the system is more complicated, as it varies continuously. We can either define 'too hot' and 'too cold', and have a heater that can be turned on and off, or we can invent a heater whose power output can also be varied continuously. In the latter case, the variety of input and output is matched; although we can't count the possible values that the ambient temperature might take, we know that they can be put into correspondence with the settings on the heater. In the former case, however, we've done something more interesting – our first piece of 'variety attenuation'.

Variety engineering for beginners

The ambient temperature of our squirrel cage could take practically any value (within a realistic range). If we make a decision to reduce our information set to 'too hot' and 'too cold', we can match it to a regulator with states of 'heater on' and 'heater off'; we've built a thermostat. Doing this isn't difficult – we just decide to throw away some of the information, on the assumption that it's not relevant.

That might end up being a bad decision, of course (if the ambient temperature rises above 100 degrees, for example, perhaps because the lab is on fire), but there is a huge saving in the amount of variety and information that we need in the regulator. This sort of decision is fundamental to the cybernetic analysis of systems; you are always attenuating variety in some way or other, unless you are describing a system that consists of everything in the universe. The simple act of drawing the black boxes is a form of attenuation, because it's a statement of what you're going to pay attention to.

Going hand in hand with variety attenuation is the concept of 'variety amplification', which is harder to understand. You can't really amplify variety; information is related to entropy, so amplifying it would put you in breach of the second law of thermodynamics. But you can do the same trick that's used in a stereo amplifier, where a small voltage is used to switch a much larger one on and off, giving the illusion that a small sound has been made bigger. The corresponding move in variety engineering is to build a regulatory subsystem within the wider regulator.

Again, the example of the thermostat might be the easiest way to understand it, if we expand it slightly. If you have one squirrel cage, you might be able to regulate the temperature yourself by looking at a thermometer and turning the heating up or down. If you have a hundred squirrels to look after, all located in different rooms, the task becomes significantly more difficult. But by making a one-time investment and fitting mechanical thermostats to each cage, you can use a different source of variety (the effect of the environment on a heat-sensitive switch) to replicate a huge number of decisions which you would have been unable to make individually.

Thinking about a thermostat also brings to mind one of the most powerful techniques of variety amplification – that of 'regulation by veto'. It works by dividing the entire potential range of temperatures into two regions, which might be defined as 'acceptable' and 'too low'. If the temperature is too low, the thermostat vetoes the current state of the system. If we pretend that instead of turning an electric heater on or off, the thermostat works by opening or closing the window, we could put it in competition with another sensor – a carbon dioxide monitor, say. If both these systems have a veto on whether the window is open or closed, the system will either settle down to a state in which both ventilation and temperature are regulated, or it will oscillate and require higher-level intervention.

Regulation by veto is a powerful way of solving huge and complicated search problems, because vetoing a situation as unacceptable is a very informationally 'cheap' action. Ross Ashby demonstrated this by building a 'homeostat' from electric circuits. The circuit was arranged such that each of four connected compass needles would effectively act as a voltmeter, either floating in the middle of its range or being dragged to one side by the local electromagnetic field. It was also arranged so that when the needle was dragged to one side, the voltage dial on its corner of the table would cycle randomly through its possible settings, meaning that it would 'veto' the current configuration of the system. If it was floating in the middle of its range, the cycling would stop, indicating that the current configuration was 'acceptable' to this needle.

All four compass needles would be doing this at the same time, until they all reached the floating position, at which point the system would be stable. Four linked inputs with thirty states each meant that the system could have 810,000 possible combinations of states. But the homeostat reached its point of stability quickly, and could always recover it when the settings were changed.

Splitting the problem

I've brought in another variety engineering trick here – and one that's almost as important to the cybernetics of organisations as amplification and attenuation. At the start of this example, I talked about treating the squirrel-plus-cage system as a single black box, but now I've begun to talk about the ambient temperature of the cage as a phenomenon on its own. Isn't this a bit of a cheat?

To an extent, yes. The technique that I've used is the separation of a big problem into a set of smaller problems. As you might recall from the Cube-turning Society, this really changes things in terms of complexity – if you smash your Rubik's Cube into twenty-six sub-cubes, you can 'solve' them all quickly. In this case, adopting this variety amplification/attenuation* method both destroys the cube and doesn't solve the problem. In the squirrel cage problem, though, for most purposes the ambient temperature can be separated from food, water and other variables. And the more variables you can do this with, the more your ability to regulate is amplified.

Furthermore, the limits on the original regulator's ability to build subsystems are only practical. To put it simply, if you can build one thermostat, you can build a thousand. This is one of the key moves in cybernetics, and one which is particularly vital to management cybernetics. When faced with unmanageable information flows in organisations, we can build systems that regulate the variety coming in, operating as much as possible through chains of cause and effect rather than through individual managers having to make decisions.

---

* In a lot of cases, the question of 'does this attenuate the variety of the environment or amplify the variety of the regulator?' is one to leave to the philosophers (see p. 67). It's a matter of cybernetics-as-description rather than cybernetics-as-rigorous-theory. The convention in management cybernetics literature, as far as I can pick up from context, is that Stafford Beer and his followers tend to talk about 'amplification' when it's really obvious that someone is building a device, implementing a management system or hiring some subordinates, and 'attenuation' most of the rest of the time.

That concept is the original decision nobody made, with the trademark cause-and-effect structure that makes it difficult to appeal against and which removes the link to an accountable decision maker. The accountability shield is a *solution* as much as it is a problem.

The squirrel cage model shows how it's possible to analyse and construct control systems by variety engineering – ensuring that the regulator has enough information-handling capacity to deal with whatever gets thrown at it. We can see here that, from our original example of a single cage with a single squirrel and a human operator, you might be able to expand to a small squirrel farm with hundreds of cages. In this farm, the original operator would be attenuating the variety of inputs by making reasonable assumptions about squirrel-husbandry policies, and amplifying their own variety by building thermostats and employing people to weigh the rodents and clean the cages.

Further complications arise as you start to consider bigger and more realistic systems. For example, you usually need to respect the fact that the system is dynamic in time, and to ensure that the regulator can take in information faster than the system can generate variety, and respond to it quickly. You also need to build translation capacity into the system, so that every black box receives its inputs in a format that it can convert to outputs. In many large organisations, a significant proportion of the staff carry out actions that might be seen as the equivalent of taking a note on the vet's prescription pad and ensuring that it affects what's on the squirrel's feeding tray.

Engineering failure and system breakdown

This is the basic building block of the cybernetics of organisations: variety engineering aimed at matching black boxes to their regulatory systems. We can see that it's a combination of art, practice and science that depends on the description of the system in terms of the boxes and the relationships between them. If that description is sufficiently matched to reality, then variety engineering will help to ensure that it can be regulated properly and show where resources need to be added, or where capacity can be saved by using processes and accountability sinks. But although variety engineering is a scientific process, the original description is subjective and fallible. For a real-world system, rather than an introductory exercise, getting to the starting point will involve its own assumptions and analogies; it will depend on whether problems that are interconnected have been split up, whether information has been thrown away which might be needed, and so on.

This is how the cyberneticians proposed to diagnose the system – to psychoanalyse the non-human intelligence. Decision-making systems break down if the variety of their environment is not matched by the variety of their means of regulation. Implicitly, a control system is a model of its environment. If the model is missing important sources of variety in the environment, overestimating the variety of part of the regulatory system, or assuming that information is being transmitted when it isn't, the system will gradually drift out of control without anyone necessarily understanding why.

The art of the thing was to come up with a sensible description of the environment, the system and its subsystems, so that it could be seen whether the principles were being respected in the connections between the components of the system, and between the system and its environment. The question of where to draw a line to demarcate 'the system' from 'its environment' often has to be solved by stipulation. Stafford Beer's answer to this problem was to develop a further set of principles to structure the process of drawing the black boxes and to give the analyst a clearer sense of what connections needed to be made. He based this partly on his own management consulting work, and partly on general deductions about what kinds of things a system would have to be capable of in order to remain regulated and orderly in a changing and chaotic world. This was his big contribution; although everything was built on Ashby's law of requisite variety, it was the 'viable system model' which underpinned his claim to be the father of management cybernetics.

5

Cybernetics Without Diagrams

We shall not, however, be much interested in any exact
measure of largeness on some particular definition;
rather, we shall refer to a relation between the system
and some definite, given, observer who is going to try to
study or control it.

W. Ross Ashby, *An Introduction to Cybernetics*, 1956

One of the most characteristic features of management cybernetics is the diagrams. Huge diagrams, with lots of arrows
overlapping each other. One of Stafford Beer's books has a
copyright-free annex at the back, where he encourages readers
to photocopy his diagrams and blow them up to A2 poster
size, before pinning them to noticeboards and writing captions
in the boxes. Another book has whole chapters in which he
shows you how the final diagram is built up from half a dozen
simpler ones, followed by a chapter in which this is shown to be
a simplified excerpt of a more complicated diagram. At various
points in books like *The Heart of Enterprise* and *Brain of the
Firm*, he bemoans the limitations of a flat sheet of paper when
it comes to showing multiple levels of connection, though it's
not obvious that three dimensions would be enough if he had
them.

This makes for heavy reading, unfortunately. Diagrams

present you with the information 'all at once' and leave you to work out the flow of cause and effect for yourself, while a verbal explanation usually presents you with the story of cause and effect and leaves you to remember the connections. In the context of systems, where feedback is ubiquitous, the relationships are vital and the flow of cause and effect has no obvious start or end point, it's not hard to see why people draw diagrams. But this emphasis on connections means that diagrams are often ineffective ways of explaining something for the first time; they give you a network of relationships, but in a context that doesn't tell you much about what the things are which the relationships are holding between.

So, I'm not going to draw a single box or arrow; I'm just going to explain, in words, the main elements of Stafford Beer's viable system model. I'm going to leave out as much detail as I can; if you're really interested in the system, there's no substitute for *Brain of the Firm* and *The Heart of Enterprise*. I won't lie; it's still going to be heavier than any other chapter in this book. But at the end of it, you'll have the toolkit, and you'll be able to understand how to apply the model to the big systems of accountability and control that have gone so wrong over the last few decades.

The trick to understanding it is to always keep sight of the basic underlying structure, which has five components. You have the part of the system that does things. There is the part that stops things getting in each other's way. You have the part that decides what to do today. There is the part of the system that's responsible for looking at how the environment is changing. And then . . . to be honest, it's difficult to explain the last part without going into more detail. Let's dive in.

The viable systems

The first part of the model is System 1, or *operations*, the bit of the organisation that is involved in making change in the real world. It's a part of the system that could, in principle, be turned into an independent organisation. In fact, it has its own internal management, which is also important to the structure of the model. Hold on to that thought – we'll come back to it.

If you set System 1 loose, though, you could expect a bit of chaos – the same squirrel being fed a dozen times while another one starves; two classes showing up at the same lecture hall and the like. Systems that aim to prevent this sort of thing happening by enforcing rules for sharing and scheduling are *regulatory*: System 2.

Stopping here for the first of several breaks, think back to the example in Chapter 2 when we were talking about 'resource bargains' as a style of management. The example used there was of a platoon of soldiers being asked to capture a bridge. If you were using Beer's model to analyse the army, that platoon would be a System 1, and it would have interacted with a number of System 2 operations – a quartermaster to assign the equipment, a battle plan to stop the platoon firing on its own troops, the system that assigned soldiers to the platoon in the first place, and so on.

And on the other end of the resource bargain, we have the battlefield commander, with the job of sending different platoons to different tasks in service of the overall military objective. Coordinating the activities of a group of different System 1 operations is the first real management function in Stafford Beer's model, and it's called System 3 – *optimisation* or *integration*. This part of the system directs the management of each individual operation in order to coordinate their activities towards a particular purpose. It's important to be clear here that this level of organisation exists because there are always several different System 1 operations to coordinate in service of a higher-level purpose. It's unusual to find a system that's only doing one thing, and if you think you've found one, it's more likely that you haven't analysed its operations carefully enough.

Keepers of the resource bargain

System 3 is where you start finding *management** jobs – those that are entirely devoted to communication and administration within the organisation. It's also where system-level account-ability is established; the key activity of the integration and optimisation function is to agree the resource bargains with the System 1 units, and to ensure that they are being kept. And consequently, it's the first place you might start to look if you think that unaccountability is creeping into the system. We'll be looking at this a lot, later.

It's easy to get confused between systems 2 and 3. Both of them look like they're doing the same thing, in making the operations accountable to one another and to the wider organ-isation. The difference is that System 2 is all about preventing clashes and managing conflicts, while System 3 is concerned with achieving a purpose. On the ground, a useful way of drawing the distinction is to look at the management functions which everyone agrees to be necessary, as opposed to those that they complain about. In consulting assignments, Beer often found that people who were System 3 managers would pretend that they were in the harmless and necessary System 2.

---

* Stafford Beer refers to the part of the system responsible for internal organisation as the 'metasystem', one of very, very many newly coined jargon words which I am going to attempt to shield you from.

Management, and exceptions

We haven't yet talked much about variety engineering, so let's look at how it takes place at this level of the organisation. The System 1 operations have to interact with the outside world, which means that they deal with a load of 'variety' every day; their information-handling capability has to, by the means we discussed in the last chapter, be matched to the range of possible things they might have to deal with. As well as the logistical problems of production, System 1 has to deal with customers, who often do things that don't fit with the priorities of the organisation. The various operations also interact, so they are sources of variety for each other too.

What happens to this variety? Most of it is matched by the intrinsic capability of the people and things in System 1; the variety of the raw materials is matched to the variety of the machinists who work with them, the variety of the customer behaviour is matched to that of the sales force, and so on. System 2 exists to absorb the variety of the interactions between operations, to the extent that this variety is predictable, constant and susceptible to the design of self-regulating systems (you might think here of a school timetable, which ensures that different teachers don't try to use the same classroom). What's left is the residual variety, the things which aren't just day-to-day decisions. System 3 has to be big enough to deal with this variety.

It's a concept that's recognisable to anyone who has hung around consultants or business school academics as 'management by exception'. That's the relatively commonsensical principle that as much as possible, people ought to be given tasks to do and left to achieve them. An 'exception', in the jargon, is an 'exception to the rule' – something the business unit comes across that it can't deal with and so 'escalates' to the level above. In a company or organisation run on these principles, the job of management consists mainly in dealing with these sorts of unusual cases, and then occasionally revising the objectives handed down to the level below.

If you believe Stafford Beer, there's a pretty strong result here in terms of battles between competing management philosophies. Management by exception is not just common sense; it's the right thing to do, objectively. The alternatives don't add up and don't respect Ashby's law of requisite variety. So they fail in one of two ways.

If the System 3 middle management tries to fight the variety generated by the System 1 operations single-handed, it's going to be overwhelmed. System 3 is a central coordinating function without the resources to handle that volume of information. Everyone who's worked in a dysfunctional organisation will recognise this as 'micromanagement' and will be familiar with the ways in which it breaks down. The environmental variety coming in from the suppliers and customers doesn't get handled at the right level because System 1 employees aren't allowed to make decisions. Meanwhile, the coordination, communication, integration and planning functions are neglected because System 3 is spending all its time trying to do someone else's job. Middle management becomes bloated and overstaffed as it tries to add variety to itself, while the operations are miserable because of constant interference from people who don't really understand what they do.

On the other hand, the central functions can be neglected by taking the opposite approach – that of excessive delegation. If everything is left to the operating-level management, there is no coordination of the resource bargains and no planning. With no adequate System 3 function, the operating units have no means to resolve their conflicts other than through internal politics, with resources allocated by grabbing and hoarding them. Nobody talks to anyone else, and the System 1 level becomes bloated with 'operational' managers who spend all their time treading on each other's toes and fighting turf wars.

It's important to note at this stage that we're talking about management *functions* here, and that these 'systems' shouldn't be expected to fit nicely on the organisation chart of a company. Apart from anything else, there are many different ways of dividing up the operations into different potentially viable System 1 functions. Plenty of organisations have no formally identified central planning department, but the integration and optimisation *function* is performed by an informal network of System 1 managers. That's perfectly possible, as long as they have made the mental leap of understanding that from time to time they need to adjust their thinking to perform a coordination role for the benefit of the organisation. Stafford Beer occasionally seemed to suggest that this kind of informal internal networking could be the best way to create System 3, which was why a big lounge at head office with whisky and cigars was important.

So if we're analysing an army at war, System 1 is the platoon, System 2 is the quartermaster and System 3 is the battlefield commander. But above this level, it's better to think of a different analogy. How would we apply the model to a different kind of organisation – say, an orchestra?

The viable orchestra

You could make a case for regarding anyone other than the musicians as the operations (System 1) of an orchestra, but it would feel contrary. The musicians fit all the criteria – they're capable of independent existence and they're responsible for interacting with the environment to deliver the system's output.

The regulatory (System 2) systems can be identified quite readily, too. The orchestra has a system for making sure that everyone is playing the right notes at the right time; depending on what kind of band it is, there's either a score determining every note, or a chart with the chords on it. That's another example of a 'resource bargain' which can be negotiated in different ways; a chamber orchestra is given all the notes to play, while the members of a jazz orchestra are required to improvise their own contributions within an overall framework.

Note, however, that there's a lot more to delivering the music than obeying the score – the musicians are responsible for playing in tune and in time, and responding to what they hear from the people around them. Another part of the resource bargain between the musicians deals with the shared supply of volume and audience attention; some sections need to play quietly so that others can be heard, some are designated as accompanists and some as soloists.

At the front is a conductor – and here we can see why Stafford Beer often said it was difficult to separate the operational systems. The conductor has one definite System 2 role: to mark the beat so everyone plays in time. That's clearly regulatory in nature, but the conductor isn't just a human metronome. They're responsible for the dynamics of the piece, directing instruments to play softer or louder at the appropriate points, and to change the tempo according to the needs of the piece. The System 3 integration and management role of the conductor is to provide the feedback to the musicians about the noise they're making and how it needs to adapt.

And an orchestra doesn't just exist when it's performing. In rehearsals, the conductor plays a role that crosses between operations and management. They're responsible not only for making sure that the musicians know the pieces and are capable of playing them, but also for making decisions about interpretation and emphasis. That's the orchestra's equivalent of the resource bargain; the musicians are accountable for delivering the conductor's vision. If they're not capable of doing that, it's part of the System 3 function to handle human resources and replace them.

So now we move on to the higher functions. I picked the example of an orchestra for this part of the explanation because there's a clear boundary between decisions made during its performances (the operations) and other kinds of decision. Conducting an orchestra in a piece of music is a particular job, but a lot of other decisions need to be made to provide the context for the performance. As well as the 'here-and-now' functions which ensure that the music is played, the orchestra needs to be able to make higher-level choices – picking the repertoire and deciding where and when the performances will happen. The conductor might be involved in this, or there might be an artistic director or a tour manager – the point is that the function needs to exist.

This is System 4 of the model, often described as the *intelligence* function. Its defining characteristic is that while System 3 manages things happening 'here-and-now', System 4 is responsible for 'there-and-then'. It is meant to be dealing with information from those parts of the environment that aren't in direct contact with System 1, and so which are capable of generating shocks that can't be handled by the 'management by exception' system. The usual reason why this part of the environment isn't detected by the operations is that it doesn't exist yet; the key job of System 4 is to make sure that the resource bargains System 3 strikes with the operations will remain feasible following anticipated future structural changes.*

It's also important to note again that systems of this sort don't necessarily match up to organisation charts, and that individuals can appear in different roles in different functions and contexts. For example, when on stage, the piano player is part of System 1 operations, and will follow the conductor's instructions. If, however, the orchestra is accompanying Elton John, the piano player will also be involved in higher levels of management and decision-making – because the piano player is Elton John!

Matching 'here and now' with 'there and then'

This brings us to the highest conceptual level of the system. We've emphasised at every stage that management is about variety engineering: making sure that every management function is matched, in terms of its information-handling capability, to the kinds of shocks and variety that might affect it. So System 3 has to be matched to the variety it manages (the exceptions and escalations coming out of System 1), and System 4 has to be matched to its own source of variety (the uncertainty about the future and things happening outside the immediate environment). But systems 3 and 4 also have to be matched to each other. This isn't particularly obvious as an engineering statement, but it's a condition of stability – and if you think about how their interaction works, you can see why.

---

* What about coping with unanticipated structural changes? We'll get there in a minute, but the honest answer might be 'sometimes they don't'. A viable system is one capable of surviving indefinitely and of adapting to unanticipated shocks – which doesn't mean that it can adapt to everything.

The intelligence and policy function (System 4) is meant to instruct the day-to-day management function (System 3) on how to reorganise itself to cope with change. There are two obvious failure modes here – changing too much, and changing too little. If the orchestra never changes its programme it will stagnate, but if it changes too often the musicians will be under-rehearsed and quality will suffer. The variety transmitted from the intelligence function needs to be matched *both* to the change it anticipates in the environment *and* to the capacity of the operational function to reorganise itself. Except by complete chance, these two are not going to be equal – so there needs to be a final management function to absorb excess variety.

System 5 is what Stafford Beer calls *philosophy* or *identity*. That might sound like an odd name to give the job of variety engineering to balance two competing management systems, but he does this because there's a specific technique that's best suited to this task, and 'identity' or 'self-creation'* is a good name for it.

It matters what kind of an orchestra this is. Having a consistent identity is a great way of reducing the variety you need to deal with, because it means that there are lot of possibilities that can be simply ignored. If this is the backing orchestra for Elton John, nobody is going to need to consider whether the repertoire should include Wagner's *Ring* cycle, or whether the tour itinerary should include a heavy metal festival. But having a consistent identity isn't just a matter of setting a kind of dogma.

In fact, understanding that identity, philosophy and purpose are tools of information management is the key to understanding the most famous slogan of management cybernetics,

---

* Beer uses the Greek word *autopoiesis* but, really, there is no need.

'POSIWID' – or 'the purpose of a system is what it does'. The identity-creating function is intrinsically linked to the variety-balancing function. In working to balance the immediate needs of the system with its response to a changing environment, System 5 is making the decisions which determine 'what it does' and, consequently, its purpose. POSIWID is not just a glib piece of cynicism; it's a description of how a system retains viability and identity. It's also, we'll see later, the key to understanding the polycrisis of the last two decades – the failures we see in the operations and management of our society have their roots in specific questions of philosophy and ideology.

The true nature of management information

The ability to translate information into action is the last piece of the puzzle, but it might have been the first piece because it's so crucial. So far, we've talked about the links between a system's parts in terms of the ability of one to absorb the variety of the other, but we need some link between this kind of information-centric view of an organisation and what it actually does.

Stafford Beer takes an approach to this problem that helps to mark the distinction between information theory and cybernetics. That solution is to say that information and action are one and the same; variety coming in from the environment, or being transferred from one system to another, only counts as 'information' if it has a causal role in decision-making. Otherwise it's just 'data' – collections of facts that hang around on disk drives, waiting to be erased* or for the format to become obsolete.

---

* There are a lot of people in Silicon Valley who might do well to consider how much money they have invested in 'data' without bearing this distinction in mind.

Stafford Beer's decision-centric definition has significant practical consequences; his model requires additional axioms beyond the general one that the informational balance sheet has to match up. For a piece of data to have the capacity to affect decisions, it has to arrive *in time* and *in the right form*.

Every channel of communication between systems not only needs to have enough bandwidth to carry the variety that it's meant to transmit, it also needs to be equipped with enough translation capacity at each end to ensure that the signal is understood. As we've noted, the signals have to be sent quickly enough and monitored regularly enough to allow them to be the basis of action, without generating oscillations. That might even mean that 'operational' management at the level of System 3 needs to work on the basis of forecasts of the near future, rather than reacting only to their immediate circumstances.

This sort of thing is perhaps easiest to see in the context of the System 2 regulatory systems. In the case of our imaginary orchestra, the majority of the variety is generated by the musicians themselves and their interactions. They have to be aware of some things about the external environment, but the key issue at the operational level is ensuring that their activities are coordinated.

The conductor's podium is a variety amplifier to make this happen; in every bar, there are four beats' worth of information that *have* to be communicated within a given period of time (there's no use receiving timing cues even a split second too late – as you'll know if you've ever tried to organise people to sing 'Happy Birthday' over a video call), and that's achieved by making sure everyone can see the movements of the baton. The musicians also need to understand the conductor's hand gestures and body language, so that they know when to play louder or softer. They also need to build up a memory of how the piece sounds, so that they are aware of what to expect. This building up of communication, transmission and translation capability happens in the rehearsal room; the key purpose of most forms of management training is to establish the channels and translation systems, so that people are able to handle variety in the real-life performance of their tasks.

Recursion, embedding and accountability

And then we reach the final part of the model, which was slightly foreshadowed when I mentioned that System 1 operations have their own internal management. It's natural to think of an orchestra as a single system, but even this was a somewhat arbitrary choice. A dance hall owner might see the orchestra as one of their System 1 operations, with a calendar and equipment as its System 2 regulator, a stage manager responsible for System 3 integration with the lighting, the sound and the drinks concessions, and a System 4 operation for booking the acts. Alternatively, if we were a management consultant hired by Elton John, we might have decided to break up the orchestra itself into component systems and think about the rhythm section having its own systems to maintain viability within the overall orchestra.

The overlapping of different systems – and the tendency of individuals to have different roles at different levels of abstraction – is a key part of Beer's theory, and one of the main reasons why his diagrams got so complicated. He claims that every 'viable system' needs to have all five of the functions described so far in order to be capable of long-term survival, but that every such system can also be seen as System 1 within a larger system. Similarly, since we defined System 1 as part of an organisation that could in principle be a viable separate organisation, the internal management of System 1 needs to have its own equivalents to systems 2, 3, 4 and 5; it needs internal regulation, optimisation and intelligence, and a balancing, identity-preserving function of its own.

Often, when you're trying to diagnose why a system is failing, you need to consider both the larger system in which it's embedded and the organisation within its operations. A great source of management problems, for example, is that organisations often fail to identify some of their operations as distinct systems, and so they lack their own internal 'higher functions'. A division of this sort will generally be a 'problem child'; unable to absorb its own environmental variability, it will bounce from crisis to crisis, taking up disproportionate time and effort on the part of the middle managers to which it has been assigned.

We might consider, for example, a string section in our imaginary orchestra where the management keep hiring a load of string players on an ad hoc basis for each performance, without recognising that it's a viable System 1. Because there are no structures for getting the violinists to talk to the cellists, they will end up taking up rehearsal time settling interpersonal and musical differences, requiring more effort and resources to reach a point where they can deliver an acceptable performance.

If the orchestra is lucky, the players will construct informal arrangements to serve as regulatory and optimisation functions. But they're unlikely to evolve a System 4 intelligence function, or a System 5 group identity to balance their immediate needs with the future, so every reorganisation or change will upset them all over again. In fact, these informal understandings are usually crucial parts of management; any organisation can create chaos for itself by having a lot of employee turnover and making sure their operations don't form viable systems.

The chain of recursion, in principle, goes all the way up to the scale of the global political economy and down to the individual human beings. This might be regarded as a weakness in the theory; it arouses suspicion that what was meant to be a straightforward technique for checking up on variety matching has ballooned into a cosmic theory of everything. But it also provides insight into understanding the 'death of accountability' in large organisations.

Red flags and red handles

If you restrict yourself to considering a single set of systems, the viable system model is fairly comprehensible. You can think of the five parts as soldiers, quartermasters, battlefield commander, reconnaissance and field marshal, or you can think of them as musicians, conductor, tour manager, artistic director and Elton John. Or whatever mnemonic helps it stick in your mind that a viable system is made up of operations, regulation, integration, intelligence and philosophy. But in a real-world organisation, there are many different levels of systems, each one embedded in the level above.

And the different levels of systems aren't necessarily aware of each other. For one thing, some of the important systems may be informal networks, with no systematic reporting or consistency of personnel. The functions of the viable system don't match up to individual responsibilities on an organisation chart. But more fundamentally, the whole *purpose* of having different levels of organisation is so that you don't need to communicate all the information from the lowest levels to the highest. The reason that these systems exist is because dealing with the variety and chaos of the world, combined with the interaction of all the parts of a large organisation, is something which generates unimaginably huge amounts of information. Most of that information has to be dealt with locally, as it arises, rather than transmitted to a central planner. This was what Stafford Beer saw as so correct in Hayek's economics.

But there's a serious potential failure mode here, which relates to the links between information, time and action. The operating principle in this worldview is that of management by exception; at every level, a system is meant to either handle an event itself or pass it on to the level above. So the bigger and more difficult a problem is, the more levels upward it gets sent, until it reaches someone or some system that can command enough resources to cope with it.

The problem is that this takes time. At every stage, information has to be collected, a decision has to be made to pass the problem upwards, and then the problem has to be communicated. And sometimes things arrive – at the lowest levels of the organisation – which are both very big, in terms of the resources needed to deal with them, and very immediate.

There is a solution to this kind of problem in the viable system model – Beer emphasises it in his later work, particularly in the parts of the second edition of *Brain of the Firm* that were written after his experience in Chile. A truly viable system needs to have communication channels that link the operations to the higher-level management functions and even to higher levels of recursion. Stafford Beer called these 'algedonic signals' – a kind of neurological metaphor, coined from the Greek words for pain and pleasure.

It might be easier, though, to understand them as the kind of messages sent by the red handle in a train driver's cab. As well as a physical check on the individual vehicle, the emergency brake sends an organisational message to the whole railway, informing it that a piece of track can't be used. An individual driver doesn't have the authority to rewrite the timetable for the railway that day – pulling the red handle is how he or she indicates to the signalling system that something is threatening catastrophic or unacceptable consequences to the viability of the system, and that it requires immediate action.

Red-handle signals are problematic for organisations. If they're being generated frequently, this is an indicator of missing, misidentified or dysfunctional systems further down the organisation. And by their nature, they go from one level of management to another, bypassing the usual systems in between.

So these signals are a channel which isn't often used, which almost always carries bad news and which cuts across levels. It's not difficult to see why they might not be allocated sufficient resources; only in contexts where they prevent the equivalent of trains crashing into each other at high speed are they respected. Cutting out the channels for red handle signals is, unfortunately, a way of constructing an accountability sink and economising on bandwidth at higher levels of management, so it tends to happen when those systems are under pressure. As we'll see in the next few chapters, this has been a big part of how we got to where we are today.

That's more or less it. I wish I could have cut more, but now we've got the essential elements of the toolkit of management cybernetics.

- There are five core functions, and if any of them are missing or under-resourced, the flow of information won't be balanced with the capacity to process it.
- Information only counts if it's being delivered in a form in which it can be translated into action, and this means that it needs to arrive quickly enough.

- Systems preserve their viability by dealing with problems as much as possible at the same level at which they arrive, but they also need to have communication channels that cross multiple levels of management, to deal with big shocks that require immediate change.

The second half of this book is going to start putting some of these tools to work. But to demonstrate how useful this kind of analytical system can be, let's quickly review one element of the biggest crisis of the last twenty years: the question of how central banks of the developed world allowed the global financial crisis of 2008 to happen.

The cybernetic history of the global financial crisis

Central banks are the institutions to which the task of controlling the monetary system is assigned. They have a variety of operations, which need to be managed, regulated and integrated – they might set interest rates, supervise the banking sector and intervene in foreign exchange markets, among other functions. The governing statute of a modern central bank is a sort of resource bargain in terms of the delegation of power over policy, in return for keeping inflation stable and keeping the monetary part of the economy functioning. The central bank is allowed independence, conditional on promising not to cause so much economic damage that it imperils the political stability of society as a whole. In other words, it's a System 1 operation of the government's economic management.

And like every operational System 1 of a larger system, central banks need to be viable systems themselves, with all the building blocks of variety-balancing subsystems, the communications channels to connect them and the red-handle warning systems that respond to threats. When that doesn't happen, things break down; conversely, when something bad has happened, you should be able to explain why the environmental variety got out of control.

During the 2000s, for example, the world's central banks thought everything was fine – they made up nicknames for the period of stability that lasted from the late 1980s to the collapse of Lehman Brothers. Alan Greenspan of the US Federal Reserve called it a 'Goldilocks economy' – not too hot, not too cold. Mervyn King of the Bank of England called them the 'NICE times', an age of non-inflationary continuous expansion. Probably the most commonly used name was the 'Great Moderation'. All the while, a huge debt bubble was building up, with house prices spiralling out of control.

What's odd is not that the housing bubble burst, or that its debt financing turned into a massive wave of bankruptcy, but that the central banks were taken by surprise. The American housing market peaked in 2006. In 2007, defaults began to grow, the investment bank Bear Stearns had to be rescued and a few of the more aggressively leveraged financial markets were shut down. But there was hardly any coordinated policy reaction until the bankruptcy of Lehman Brothers on 15 September 2008.

This ought to immediately alert us – the problem was in System 4, the 'there and then' intelligence function. Absence or weakness in this system is one of the most common problems in organisations, and central banks have a few organisational quirks that make them particularly vulnerable. Their aim is to keep a small number of target variables within a relatively narrow range, while also ensuring that there isn't catastrophic breakdown in the system as a whole. That requires a System 3 to implement and integrate those actions of the central bank which involve intervening in markets and making regulations, and a system that looks ahead to see whether structural changes in the economy might require a reorganisation of those policy actions.

It looks like that's what central banks do: make forecasts of the future, publish them and update them as the situation changes. But forecasts are their version of 'here and now'; when you hear them talking about inflation targeting, they mean inflation *forecast* targeting. Interest rate changes affect the economy with a lag, so they need to be implemented on the basis of the expected path over a period of time – not just because of what's happening on the evening news.

If something isn't in the forecast, it isn't part of the information set and can't affect the decisions – which is why central banks try to supplement their economic and statistical models with other data sources. But there's a more subtle problem – if you're doing all this forecasting, it's hard to believe anyone who tells you that you're not looking to the future.

A real System 4, though, is explicitly concentrated on those parts of the environment that *aren't* yet relevant to what it's doing. This capability was weak in the central banks; they were not looking for things that might have upset their policy-making framework. The information was there, but it hadn't been organised into the decision-making process and didn't shape the view at the management or operational levels. It remained as mere 'data' or was attenuated away by simply ignoring it: the 'information-processing system of last resort'. Only when a red handle had to be pulled – what Stafford Beer might have called the 'algedonic squeal' of the Lehman Brothers bankruptcy – was the huge capacity implicit in the central bank's powers brought to bear.

In the years when it looked like things were going well, the central banks developed a view of the world in which the changing structures of global finance weren't part of their job. They didn't pay attention to the debt bubble, and they had got rid of the communication channels that might have carried the red-alert warnings up to the highest levels of policy-making. More precisely, they had got rid of the *translation* systems. There was no shortage of people warning that there was a problem in 2006 and 2007, but none of these warnings was given in a form that could be recognised by the world's central bank governors as requiring action. A failure to build System 4 and balance its variety against System 3 is itself a failure of System 5 – that's the function that has the job of balancing 'here and now' against 'there and then'. So, using the viable system model to diagnose the causes of the global financial crisis, we end up with a rather interesting conclusion. Where things went wrong was a matter of *philosophy*. The central banks had an identity-creating function, but it had failed; it defined their purpose in such a way that they failed to understand that particular kinds of information were relevant to them.

And the particular philosophy and identity which had taken over the central banks was a big part of the story of how things have progressively gone wrong since the 1970s. What we're talking about here is something that needs to be looked at in a lot more detail. It's the rise of the economists, combined with the crisis of the managers.

INTERMISSION

Decerebrate Cats

Stafford Beer had a few favourite startling metaphors that he'd drop into lectures or book chapters to wake you up. In order to make a particular point about the importance of systems 4 and 5, he'd occasionally compare an organisation to one of the animal victims of early neurological experiments.

If you cut the connection between a cat's cerebellum and the rest of its brain, it still looks alive. It can walk, right itself when it falls over, and eat and drink when food and water are placed in front of it. It might even be able to groom and clean itself, but it is no longer capable of purposive action. The decerebrate cat will survive as long as its environment is compatible with this, but it can't generate its own responses to change – it needs to be sheltered from the everyday world, using other resources. Beer made the point that like the cat, you will often see a 'decerebrate organisation'. (His typical example was a university.)

An organisation in this situation is one that has, for one reason or another, stopped paying attention to some kinds of information. It's only aware of its immediate surroundings – this quarter's revenue, the current staffing level, things like that – and has lost the ability to make plans. Like the cat, it can continue to survive as long as nothing major changes, but the next time it encounters a shock, it will go into crisis. In that crisis, if it doesn't find resources from somewhere to re-establish the missing functions it will simply collapse.

I've worked in a few organisations of this sort. I've also been in the even more terrifying situation of working for a company which is actively in the process of decerebrating itself. When you see this happening, update your resume and start calling recruiters.

PART THREE

THE BLIND SPOTS

6

Economics and How It Got That Way

Economics is a science of thinking in terms of models,
joined to the art of choosing models which are relevant
to the contemporary world.

John Maynard Keynes, letter to Roy Harrod, 1938

To the extent that any of it was designed at all, this world was
designed by economists. Of all the social sciences, economics
was the one that embedded itself in the governance and regulation of public life – the higher functions of the system, the ones
that balance future and present needs.

And as with any other identity-creating function, economics has been a major engine of information attenuation for
the control system. Adopting the economic mode of thinking
reduces the cognitive demands placed on our ruling classes
by telling them that there are lots of things they don't have
to bother thinking about. The adoption of economic growth
and efficiency as a core philosophy and cost–benefit analysis
as a method of governance means not only that thousands of
possible policies can be rejected without serious consideration,
but also that whole approaches to human life never need to be
considered.

Before we really start using the cybernetic toolkit to diagnose the crises of the industrial world, it's worth spending a

bit of time thinking about this widely misunderstood but highly influential thing called economics. It has blind spots and inconsistencies, which often lead to predictable flaws in our institutions. When non-human decision-making systems become pathological in society today – most characteristically, when decisions are made which have disastrous long-term consequences as a result of relatively trivial short-term cash savings – the pathology is often directly related to something that seemed like a good idea to an economist.

Among the economists

I am an economist myself, of a sort. This doesn't just mean 'someone with an economics degree' or 'someone who earns money from the study of economic phenomena', although I have, and I do. Being an economist has been called a style of rhetoric and a way of thinking as much as an academic discipline. I've joked in the past that the 'ist' at the end of 'economist' is the ist of ideology and not of science – analogous to 'Trotskyist' rather than to 'meteorologist'. Economists variously describe economics as a subject, a science, a discipline and even a profession, but it's really a commitment to certain ways of modelling the world.

Once upon a time, economics was the study of the economy – like geology is the study of rocks – but the last seventy years has seen the eclipse of what used to be called 'institutional economics'. Back in the 1930s and 1940s, to be an economist was to make detailed studies of particular economic institutions, compile statistics about them and then publish them, after which they would sit on a shelf for decades, in the hope that someone might come along who wanted to know a lot about rayon, or boric acid, or shipping in the South Pacific.

This has been all but eclipsed by another kind of economics. One of Friedrich Hayek's contemporaries defined economics as 'the science which studies human behaviour as a relationship between ends and scarce means which have alternative uses'. It's this idea of a generalised study of human behaviour under conditions of scarcity that has been responsible for the economists' intellectual imperialism. Because if you announce that you're the experts on human beings when they have to choose between different priorities under conditions of scarce resources, well, when *aren't* resources scarce? When *don't* people have to choose between different things they want?

Triumph of the optimisers

This second kind of economics became dominant after the Second World War. The same war work that spawned the Macy conferences and the beginnings of management cybernetics had a somewhat paradoxical effect on economics. Economists of all political kinds were thrown into planning jobs of various kinds during the war – from tasks like surveying the capacity of the USA to produce aluminium, to abstract constrained optimisation problems in areas from food rationing to minesweeping. Many of these tasks brought them into the same interdisciplinary teams and 'operations research' departments that spawned cybernetics.

However, the effect on the economists seems to have been different. The profession benefited greatly from contact with pure mathematicians and engineers, many of whom greatly raised the level of mathematical sophistication in economics. But it never seemed to be broadened as other subjects were thanks to the experience of war; there was a great leap forward in terms of technique and analytical rigour, but not much else.

There might have been a sociological explanation for this – as economics was the most prestigious of the social sciences, economists had more to lose and less to gain from contact with outsiders. There were also concerns of sheer intellectual bandwidth and capacity, because the subject itself was going through an explosion of major works. For economists, during the period in which the Macy conferences were taking place and the rest of the world getting interested in cybernetics, there was a hell of a lot of to keep up with in the literature. Added to which, cybernetics didn't look too interesting to economists, at least on the face of it – as we've noted before, much of it looks like the socialist planning debate, which was regarded as a settled question.

By the end of the post-war intellectual upheavals, though, economics had changed into a very mathematical science, based on a general toolkit of 'constrained optimisation'. Broadly speaking, if you could convert your problem into one with a single numerical index of success or failure, economists would regard it as their job to help you. 'Help', in this context, meant that they would find a way to describe what you did in terms of a system of equations which related the quantities under your control and the limits on your resources to the numerical good-or-bad index, then solve that system of equations to maximise success or minimise failure.

Often – though not always – it was convenient to express the index of success or failure in terms of an amount of money. Money's useful property is that a lot of things are expressed in terms of it,* including many of the things that need to go

---

* As the comedian David Mitchell put it, there's no need for wine critics; restaurants already tell you how nice the wine is on the menu, in money.

into the systems of equations that economists want to solve in order to optimise the system – commodity inputs, hours of labour and so on. Most things have a market price that you can either look up in a catalogue or find out by requesting quotes from a few manufacturers. Even things that aren't traded on a market can be given a monetary value either by asking people what they might be prepared to pay for them, drawing rough analogies with other things, seeing how much they're insured for against loss or just guessing. Over time, the method of optimisation became more or less hegemonic as a way of dealing with questions of production, consumption and trade, and with it the culture of 'thinking like an economist'.

Economists also learned how to deal with data. Another application of the mathematics of constrained optimisation is that you can define your index of success or failure as something like the 'distance between what my model predicts and this set of observations', and tweak the parameters of your model to minimise that. A great deal of the apparatus of modern statistical analysis was invented in economics departments – and economists are often much better at it than many people working in the 'hard' sciences.

The combination of the two mathematical toolkits was extremely powerful, in political and social terms. Economics could make a model suggesting that a given outcome was optimal, before gathering data about it and estimating how far reality was from the theoretical optimum. Depending on political preferences and ideological commitments, they could then either go back and make a new model, or declare that the distance represented a failure on the part of reality and suggest a policy to bring the world into line. It's not hard to see why these people became influential advisors.

Prices of everything

The two techniques were often controversial. One of the things that got economists a bad reputation was an excess of enthusiasm for applying their methods to things people really cared about.

The fact is, large-scale industrialisation has serious costs as well as its great benefits. Pollution causes people to die; large mining projects permanently destroy the natural environment. In order to be included in an optimisation calculation, these things had to be given monetary values. Unless handled carefully, putting a price on human life or natural beauty can come across as a bit sociopathic. At the other end of the scale, lots of people believe that art and culture has a value of its own; funding models which only credit an opera house or an art gallery with a contribution equal to its ticket sales will often seem wrong to people who might not consume the product, but feel like it's important it should exist.

But within their own system, the economists had a defence. When you refuse to put a price on something, you're refusing to admit the existence of any trade-off between one set of priorities and another. As a method, 'because I say so' is neither intellectually defensible nor democratically accountable; if the trade-off is there, it doesn't disappear just because you refuse to recognise it. Like estate agents and divorce lawyers, a lot of the unpopularity of economists is thanks to transferred guilt; they are people who deal with unpleasant realities which we'd prefer not to face, acting as accountability sinks for difficult and unavoidable decisions.

That isn't what's wrong with economics. Or at least, if we're going to blame economists for failures in the world, we need to blame them for practical deficiencies in the way that they create systems, not alleged moral deficiencies in economists. And we should probably be a bit careful with our criticisms; after all, whether or not you want to admit it, if you have some inputs and you want some outputs, then you're doing economics. It might feel nice to say that not everything can be measured, but if you can't measure something, how are you going to know whether it's changed? And if you aren't interested in whether it's changed, how can you really claim that you care about it? One of the things that you must bear in mind as a critic of economics is that the modern industrial world does, broadly speaking, *work* better than any of the available alternatives, and that economists have to take a lot of the credit as well as the blame.

The intellectual vices and foibles of economists are not random deviations; they're assumptions and strategies which are used to turn an impossibly complicated aspect of human society into a subject that you can say something comprehensible about. Viewed in this way, the high-level philosophy and methodology of economics is a form of variety engineering. It's the way in which economists attenuate the complexity of the real world and amplify the power of their own models. And they got there first – economics, in its modern form, is the planning and management discipline that cybernetics sought to be.

But there are vices and foibles. Because economics tried to solve the problems of information, management and organisation before they had been rigorously formulated by Wiener, Shannon, Ashby and Beer, there are some methodological blind spots. Once you're aware of them, you can see how they cause recognisable weak points in the systems and institutions that have been designed by economists or in line with economic principles. As a subject, economics seems to have a fear and disgust of thinking about philosophy and methodology that might be described as Freudian. While other social scientists' obsession with minute discussions of their methods and rhetoric, standards of proof and what they hope to achieve might be thought of as pathological in another way, the economists' determination to sideline methodological discussion has observable bad effects. The vast majority of economists literally don't know what they're doing.

Solving it in the model

Let's take an actual example and do some economics. Consider a hypothetical country (let's call it 'England') where it takes 100 hours' work to produce a bolt of cloth, and 120 hours to produce a barrel of wine. Consider another country (call it 'Portugal') where thanks to advanced technology and a favourable climate it takes 90 hours' work to make the cloth and 80 to make the wine. Both countries want cloth and wine. How much cloth and how much wine should Portugal produce?

If we presume Portugal is a lot smaller than England, the answer is clear. Even though it can produce cloth much more cheaply than England can, it should produce *only* wine and get all its cloth by trade. You can see this by calculating the relative prices. Without trade, a bolt of cloth in Portugal would cost nine-eighths of a barrel of wine (assuming that labour is the only input). By similar reasoning, a bolt of cloth in England would be worth five-sixths of a barrel of wine. Cloth is worth more than wine in Portugal, but less than wine in England.

And that means that it's possible to take advantage of the different prices. Rather than spending 90 hours making a bolt of cloth, if Portugal makes a barrel of wine with 80 hours of labour, it can swap it for more than a bolt of cloth. Even though Portugal has an *absolute* advantage in that it is a more efficient producer of both goods, the gains it can make from trade are determined by its *comparative* advantage.

This result is one of the crown jewels of economics – the theory of comparative advantage, first set out in David Ricardo's *Principles of Political Economy and Taxation* in 1817. When the economics Nobel laureate Paul Samuelson was challenged to name a single proposition in the social sciences which was indisputably true but non-trivial, this was the only one that he could come up with.* He said, 'That it is logically true need not be argued before a mathematician; that it is not trivial is attested by the thousands of important and intelligent men who have never been able to grasp the doctrine for themselves or to believe it after it was explained to them.'

This is also a good example of one of the big blind spots of the economic method of reasoning. This one was identified by Joseph Schumpeter as 'the Ricardian Vice' in his history of economics – both in tribute to Ricardo and in identification of the problematic side of his legacy. Ever since Ricardo, economists have had a strong tendency to:

a) make a model of some feature of the economy, stripping away nearly all the complexity;
b) make a lot of simplifying assumptions, often questionable in terms of their empirical relevance;
c) show that their conclusion follows from their assumptions, which ought to be quite easy if they've made the assumptions strong enough;

---

* According to his presidential address to the Third Congress of the International Economic Association in 1969, 'The Way of an Economist'. Stanislaw Ulam, the mathematician and nuclear bomb scientist, asked the question.

d) act as if their conclusion has now been proved in the real world.

Economists do this all the time. A very influential model* of racial and gender discrimination holds that it could not happen in a perfectly competitive environment (because a company which refused to hire certain kinds of workers or serve some customers would be missing sales and not recruiting the most productive employees, and therefore would be likely to go out of business). Consequently, plenty of economists still think that making markets more competitive is a viable substitute for anti-discrimination laws. It works in the model, as long as you throw away any information that might be relevant to the actual problem.

Samuelson was correct that plenty of important and intelligent men (and women) have not been able to grasp the doctrine of Ricardian comparative advantage. But this isn't because they're incapable of understanding that price differences between countries can be used as if they were a way of converting one commodity into another. What people have a problem with is the step between the abstract argument and the suggestion that it's describing international trade.

Think of the different steps you could add to make the wine/cloth model more realistic as a description of the real world. You could include more commodities; you could stop pretending that there was only one input; you could consider brand premiums and quality. You could introduce the fact that

---

* Attributable to Nobel Laureate Gary Becker. Becker's work is a fertile source of this kind of thinking; he also wrote that because criminals were rational actors who used cost–benefit analysis, you could save on policing costs by making the punishments really harsh.

goods are made by companies. You could think about investment in production processes and long-term strategic planning.
You could ask what happens when the demand for a product in
one country affects the overall global price.

If you looked at what actually happens, you'd see that
countries often import similar kinds of goods to the ones they
export, which doesn't seem consistent with this model. So you
might decide to model product differentiation, or government
policy, or strategic behaviour. Depending on what you chose,
how you set up the model and what assumptions you made, you
could prove your own result, or the one that your boss wanted
you to prove. But which of these models would give the objectively 'right' answer? The model won't tell you, that's for sure.

Proving things 'in the model' and then acting as if they're
true in the world is a terrible habit of economists. Not much
has changed since Ricardo's day, except that the mathematics
has got more difficult; as a result, it is significantly more difficult to tell whether an assumption is merely a technical one to
simplify the calculation, or whether it is equivalent to assuming
the conclusion. As a rule of thumb, whenever you hear an economist say 'however you look at the numbers' you should think
of it as meaning 'however *I* look at *these* numbers'.

Even the assumption that there has to be a conclusion at all
can be problematic. After all, reality doesn't usually give you
straightforward answers, so why should a model? A whole lot
of economic knowledge taught at postgraduate level under the
heading 'modelling strategy' is to do with the kinds of mathematical statement that need to be put into a model in order to
make it tractable,* not because they have any relevance to the

---

* Tractability used to be easy to spot, because assumptions and
modelling decisions would be made out in the open in order to ensure

actual economy. Modelling strategy is all about making sure that you will end up with a specific conclusion rather than a cluster of ambiguous statements.

And that's important in economics, because as we said earlier, it's a style of rhetoric and a way of thinking as well as a social science, and that style of rhetoric is built around always giving an answer. Individual economists might hedge their bets as much as any other professional advisor, but the appeal of economics as a means of reasoning has always been the straightforward logical line from justifiable assumptions to actionable conclusions. Schumpeter's Ricardian Vice is a necessary one in order to preserve this feature. Economists call it 'rigour', but the application of the rigorous model to a messy real-life economic situation brings back all the metaphor and analogy, in a way which obscures what's happened.

This is one side of the Ricardian Vice, which affects economists because of their privileged position in the political and governance structures of society – a position that's dependent on their ability to give advice that looks objective and act as constructors of accountability sinks. If sociologists or cyberneticians had taken this role, we might be talking about the Durkheimian Vice or even the Beerian one. The choosing of simplified assumptions, the stripping away of relevant detail isn't necessarily problematic in itself – that's just what it means to make a model. But another aspect of Ricardianism is more particular to economists.

---

that the equations had a definite solution that could be worked out with pencil and paper. Since computer simulations became more technically feasible and professionally respectable, there is more work done in economics where there isn't a single analytic solution. But 'modelling strategy' is still there in the background nearly all the time – the central commitment to give an answer is always there.

In terms of our two visions of problem-solving from Chapter 4, economists are Rubik's Cube people; they deal with things that have interactions and connections, and those connections multiply up rather than adding up. They have developed their own ways to attenuate and cut down that variety, and they're quite committed to using them – even when it means losing a lot of important information.

Homo economicus was a friend of mine

Consider, for example, the iconic image of economics – the diagonal crossed lines of the supply and demand curve diagram, with price on the vertical axis and quantity on the horizontal one. The line representing demand slopes downwards, because customers want less of the product as the price rises; the line representing supply slopes up, because companies want to sell more of it at a higher price. You can show undergraduates a few interesting things with this kind of graph – the effects of price regulation, changing technology and so on – but let's think about the assumptions that had to be made to get there.

There are comparatively few industries where the textbook supply and demand graph is reasonably defensible as a description of how decisions on price and quantity are made. Let's consider a best-case example, in the industry of electricity generation.

Electricity is a commodity – you can't sell higher-quality branded electrons – so you can be confident that a demand curve will have the right slope. As the price rises, people will substitute other energy sources, or reduce their total consumption, and the demand for electricity will fall. But what about supply?

In this industry, economists can be sure that their theory is right – their concept of the 'marginal cost curve' is the same as one that was independently invented by electrical engineers. The marginal cost is the cost of producing one more unit of output – no overheads, no fixed costs, only the direct cost of producing that specific increment. So if you're running an electric power grid with a number of different stations, it's the cost of producing an additional kilowatt at your cheapest operating generator.

As you produce more electricity, the efficiency of that station might get worse, until it's cheaper to generate the next kilowatt increment with a different fuel source. At any level of output, though, there will be a known value for the cost of producing the next increment; an engineer might call this a 'load balancing table', but the concept is the same as the economic one.

And in this industry at least, the marginal cost curve determines the supply curve. If the cost of producing another unit is more than the price you can sell it for, then you'll lose money. If the cost of producing another unit is less than the price, then do so – otherwise you're leaving money on the table. The amount that you supply at any given price is determined by the point on the marginal cost curve where the cost of the marginal kilowatt hour is equal to the revenue from selling it – and so the supply curve for electricity generation ought to look like a slightly bumpy version of the one in the textbooks.

But can you say the same things about markets in general? Can you even say it about the electricity generation industry all the time? Of course, the answer is no, or at least not necessarily, because the analysis here depends on a number of features of this particular market situation which happened to coincide with common modelling strategy decisions made by economists.

The first such feature is that electricity is a commodity. The question of what to supply only requires an answer in the form of a quantity of kilowatt hours and a price. Most goods have all sorts of characteristics that are relevant to both the demand and the marginal cost of producing them. Even electricity isn't really like this; a kilowatt hour supplied at midnight isn't the same as one supplied at peak time. You can deal with this in a fairly straightforward way, of course, and the electrical engineers worked this one out too; just treat different hours of the day as different markets. But the further you get from a commodity good, the harder it becomes to justify this approach.

Values and uncertainties

Related to this is a second feature of the electricity generation industry: what we've modelled here is a 'spot' market. The supply and demand curves describe the market for electricity being generated and consumed right now, during a single short period. In an hour's time there will be a new market, with a different demand curve and potentially different marginal costs, and so on.

Having gone through a crash course in cybernetics, the warning lights in our minds should be flashing right now. Are these spot markets independent of each other, or are they related? Is this a long line of dice to be turned the right way up, or a Rubik's Cube?

The answer depends on the industry. If we were thinking about designer handbags, it would be more obvious – if you sell a million units at marginal cost today from a bin at Costco, that will obviously affect your ability to sell the same product at a premium price tomorrow. Electricity might be easier to model as an independent sequence of spot markets, but there will still be decisions to take that will affect its demand curves and costs in the long term.

For example, at some point the utility company is going to have to decide whether to invest in green energy. During a heatwave or a cold snap, it is going to have to decide whether to raise its prices to levels that cause hardship for its customers. It is likely to have a regulator to deal with, so it will need to consider the political consequences of some of its decisions. In the real world, dealing with these issues is the business of management. You don't make decisions based purely on costs and revenues; you always take into account your long-term brand, your regulatory relationships and your corporate sustainability strategy.

But none of this is open to economic modelling. To get a single answer from the maths, and so to preserve the ability of economics to give definite and unique solutions, you need to be maximising a single quantity. If the commodity is defined solely in terms of quantity and price, there's a straightforward index to optimise; if you know the cost function, you can calculate the profit.

If the commodity is defined in terms of a whole list of specifications, though, it's necessary to boil them down to their impact on cost and price; the equations of constrained optimisation oblige us to pretend that every corporate value or priority is really a complicated expression of beliefs about profits. So when the electricity company announces that it's keeping prices down to help customers in a cold winter, it might really be saying that this is the best way to preserve long-term demand and avoid having a price cap imposed by the regulator. The economic approach requires all other kinds of information to be translated into expectations about profitability in an uncertain future. And that uncertain future is itself not handled very well.

Time and uncertainty

There's a curious timeless quality to mathematical economics; it effectively turns the massive space of possible futures into an even more massive space of conditional plans. It then uses the maths of optimisation to pick the future path with the greatest positive – or smallest negative – value. Doing this in a mathematically tractable way requires a number of compromises, again in the name of 'modelling strategy'.

First, you end up being restricted in the kinds of uncertainty you can consider. The calculations involved are basically weighted averages; you calculate the 'expected value' of a course of action by multiplying the payoff from every possible outcome by the probability of that outcome happening. But these probabilities are not provided by nature; you have to make assumptions about them. And averaging over a probability distribution is tricky; in order to get a well-defined answer, economists generally end up making some quite questionable assumptions.

Only some kinds of uncertainty correspond to probabilities which are computationally well behaved. If you want to calculate stable expected values, things can't get too wild. Nassim Nicholas Taleb wrote a whole book about this problem; *The Black Swan* was for the most part greeted with nervous laughter by economists, because it asks a question that's very hard to answer: what *is* the risk of something like the Russian Revolution, or Covid-19, or a climate catastrophe? How *do* you put a percentage on it and include it as part of your weighted average? And when you think about it, isn't that a strange way to go about decision-making?

However, this is arguably not even the most serious problem economists face with respect to fitting time and uncertainty into a framework that wasn't really designed for it. There's also the familiar problem of the tendency of combinations to multiply rather than adding up. Because, as a decision maker considering a long-term plan, one of the basic things you need to consider is that *all your competitors will be doing the same thing*. Electricity generation is unusual in this respect; entry to the market is regulated and takes a long time, so there are comparatively few genuine strategic decisions to take. In more typical industries, your decisions are much more conditional on other people's, and theirs on yours, and on each other's, in a feedback loop that can quickly get out of hand.

One branch of economics tries to address this problem head-on. Game theory can work where the number of parties to take into account is limited, their options are similarly limited and their behaviour generally easy to predict. How many situations satisfy these criteria? Not many, but there are some economically significant ones.

For example, consider the case of mobile phone spectrum auctions. Periodically, governments give telecoms companies the opportunity to bid for licences to use parts of the electromagnetic spectrum in an auction. There are only a small number of parties that are able to participate in the auction, the spectrum only has one use, and so the price paid mainly depends on each party's perception of what each of the others might be willing to bid. This is almost an ideal case for game theory, and indeed, everyone involved in a spectrum auction tends to hire professional advisors who are game theorists. That actually makes the task even easier, because it increases the likelihood that everyone will behave according to the theory. Several governments have taken advantage of this fact to redesign the terms of the auctions so as to maximise the price paid.

Outside that specialist field, economists tend to deal with the problem of strategic uncertainty by pretending it doesn't exist. And this is how we get *Homo economicus*, a theoretical creation who not only cares about nothing except price and quantity, but who is also perfectly informed about future supply and demand schedules. If there are other parties to deal with, he can read their minds; it's assumed that everyone has perfect knowledge of the correct economic model and is able to solve it for the optimal equilibrium.

Often, though, economists don't even bother to acknowledge the existence of multiple agents. Unless they've been specifically asked to do otherwise, they are likely to make a model of the behaviour of a 'representative individual', by assuming all the preferences of every individual are identical, or aggregating all producers into a 'representative firm'.

As I said earlier, economists don't do methodology. If they did, they'd recognise that what began as a modelling strategy has become a factual assertion about the world. The purpose of all these assumptions is to tame the problem, to make it possible to treat complicated long-term problems as a series of independent spot markets.

Places where economics is absent

These features of modelling strategy generate detectable blind spots. The concentration on single-value maximisation, for example, is one of the reasons why there is no particularly useful theory of advertising. Advertising is a big industry, and an important activity of firms, and yet economics not only has hardly anything to say about it, it has no real explanation of why it exists at all.* Advertising and marketing, despite their

---

* I'm putting this in a footnote, because I'm embarrassed to type it; a popular theory of advertising, developed by Gary Becker, is that

obvious importance to businesses in the real world, produce and use a different kind of information from the price and quantity signals which the model can handle.

For a related reason, economics has left a great deal of the thinking about management and strategy to business schools. As we have seen, maximisation problems become unreasonably complicated when you allow for even a little bit of strategic interaction. The correct way to deal with this complexity, as Beer and Ashby would have told us, is to treat the system as a black box and analyse its actual behaviour. The commitment to assumptions of rationality and maximisation, though, leads to a commitment to box-opening.

It's called 'microfoundations': the idea that every large-scale phenomenon has to be seen as the cumulative result of individual decisions. As we said earlier, when economics took its turn into mathematics and physics, it adopted individual preferences as its equivalent of the basic particles. Something isn't regarded as 'rigorous' in economics if it can't be justified in terms of the optimising behaviour of an individual. There's not much room for behaviour that's based on rules of thumb, habits from the past or passing fashions; these are 'ad hoc', which is a grave insult to an economist. As a style of rhetoric and a way of thinking, economics needs to pretend that nothing is real except price and quantity; everything else is just

---

spending a lot of money on advertising signals to the consumer that the product must be good, because it generates enough free cashflow to support a big advertising budget. The competing theory can be more or less summarised as, 'People like seeing adverts for the things they buy, so they consume the adverts as part of the overall experience.' While less ridiculous, this does leave one with a strong sense that economics hasn't advanced the ball very far.

a convenient way of talking about extremely complicated sets of conditional expectations.

Here's a story about past strategic decisions: Nokia once made rubber out of dandelions. They were actually very good at it; a Finnish Air Force plane which crashed into a lake in Karelia during the war in 1942 was recovered in 1998 with its tyres still inflated. Because they made rubber, they made insulation for electric cables. Because of that, they started making cables themselves. Because of *that*, they got into the telecoms industry;* for a while they were the world's biggest manufacturer of mobile phones and today they make complicated switching equipment. An economist is committed, methodologically, to saying that all the hundreds of decisions that needed to be made to take a company along that sort of trajectory are basically identical solutions to the same optimisation problem with different numbers plugged in.

More generally, in real management situations, price is almost never a simple question of scarcity and preference, supply and demand. Pricing decisions are strategic; they are used to position a brand, to send a message to competitors, to build a relationship with a supplier or to calm down a politician. Prices are raised and lowered as part of multi-year plans, which include advertising campaigns, new product launches and even corporate mergers and acquisitions. To model them in the economic framework would require all of the apparatus of game theory, and would quickly spiral out of control. Pricing problems which might be considered too easy for the

---

* I confess – this is an oversimplification. Nokia did dozens of other things and it's very hard to trace the timing and logic of its entry into various different industries. Basically the only principle I could extract from its corporate history is that if Finns needed something, Nokia would usually have a go at making it.

brand managers' graduate training programme at Procter and Gamble would be way beyond hope of computational tractability if expressed as equations.

Markets as computing fabric

How do they get away with this? To a certain extent, they don't – people have been laughing at economists for decades about the unreality of their assumptions. But as I suggested earlier, the nature of modelling means that assumptions have to be unrealistic. We might rephrase the question as, 'How do they sleep at night?' How do you convince yourself that a model in which everyone behaves rationally to optimise their long-term interests by choosing price and quantity is a good simplification, ignoring the evidence in front of you?

The answer, surprisingly, is related to Stafford Beer's computing pond. Whenever there's a lacuna in an economic explanation, it's possible to argue that the market will sort it out. In the 1920s, the debate was won as to the superiority of markets over central planning, but this victory seems to grow in the retelling. The market isn't just a way of economising on the need to process and transmit information, for modern economists – it's a magic calculating machine that efficiently reaches the best possible solution. The dream of 'computing fabric' and of artificial intelligence has always been to get more out of the system than you put in; the dream of market supremacy is for exactly that kind of self-organising system. Rather than explaining how the agents in a model might be able to acquire their perfect information, one can just suggest that the information processing happens in the market itself.

This assumption seems to come from at least four different intellectual traditions. There's the original 'invisible hand' of Adam Smith, the idea that free exchange leads to the best outcome for all. This was given a specific interpretation by Hayek and the Austrian School in their response to the socialist calculation debate, elevating the market as the supreme method of processing information. The post-war mathematical optimisers proved that it was possible to show that an abstract representation of a market equilibrium, for instance, had some properties which made it difficult to improve upon. And from an entirely different direction, the concept of 'market efficiency' in financial economics predicted theoretically, and then established empirically, that it was very difficult to beat the stock market and so in some sense, the market was more 'intelligent' even than investment bankers.

Together, these things contributed to a sense that the market economy was a kind of computing pond; it organically grew optimal solutions to problems and, like a true artificial intelligence, you got out more than you put in. It's somewhat ironic that economics, the discipline that's meant to tell you that 'there's no such thing as a free lunch', ended up believing so strongly in an informational version of the same thing. But the idea really began to take hold as time went on; the calculation of 'implied probabilities' even took the place of conventional economic forecasting, and 'prediction markets' were promoted as ways to take policy decisions without any individual being responsible.

When we're thinking about how the modern system took leave of accountability, the assumption that the market knows best is a big part of the story. It has its roots in the way that economists dealt with a need to ignore problems of time and uncertainty, itself rooted in the need to make optimisation calculations tractable, which in turn was a result of the unique social and political position of economics.

The blind spots

This is how the cybernetic problems of the post-war industrial state have been shaped by economics. There are three major blind spots: the belief in markets as computing fabric, the flattening of time and uncertainty and the commitment to Ricardianism.

The belief in markets as computing fabric is such an obvious source of blind spots as to hardly require explanation. Faith in the power of free markets to solve optimisation problems is so great as to support the reverse inference of economists that if something appears to be a market equilibrium, then it must be an optimal solution. This might not have been too bad if it had been treated as a first presumption, but it became ideological. A great deal of the reasons why policymakers make mistakes comes from a reluctance to second-guess market outcomes. Think of housing bubbles, for example, or financial deregulation.

'Market outcomes' are a huge accountability sink, but there's also a philosophical blind spot. Policymakers and their advisors have internalised the belief that the market is a greater intelligence than anything humanity can offer. The combination of political convenience and intellectual rationalisation is a powerful one; when a possible market solution presents itself, it has a huge advantage in consideration against alternatives, whether or not it deserves one.

The damage caused by assumptions about time and uncertainty is more subtle. *Homo economicus* and the profit-maximising firm are the economists' equivalents of the cyberneticians' black box: the fundamental components of the model and the system. But they're constrained to have only specific categories of inputs (price and quantity), which relate to the output in a specific way (optimisation). Considered as a cybernetic system, an economic model is one that has arbitrarily thrown away most of its information.

When economists design systems, their weakened concept of a black box shows up in the lack of concern with the capacity of communication channels to carry the required information. Equally importantly, the problem of translating information into actionable form hardly exists; when the information set is reduced to price and quantity, and when decision-making is just optimisation of a single value expressed in money terms, there's no way to talk about ambiguity, perception or incompatible values. This is how cost–benefit analysis goes wrong; a statement like 'this passenger aircraft has a faulty stall sensor which might cause it to crash' fundamentally *isn't* a piece of information about expected costs – even if you can get people happy with the idea of assigning a financial value to human life.

And finally, there's Ricardianism. The Ricardian Vice, you'll recall, refers to the practice of solving things in the model before applying the solution to the real world. This was the reason central banks missed a generational financial crisis. It didn't exist in their model, and the Ricardian Vice, in Stafford Beer's terms, translates to a weak System 4, which is meant to pick up information about things that could change the operating environment in such a way as to render the current model insufficiently realistic.*

---

* Some economists in my experience have tried to give a partial defence of the Ricardian Vice by quoting the statistician George Box, who said, 'All models are wrong but some are useful.' As a defence of modelling in general, this is a pretty smart thing to say, but note that if this is what you really believed, you'd probably try to create a systematic process for checking and updating your models to see whether they were becoming less useful.

Ricardian Vice thinking also tends to dull the response to red-handle signals and screams of pain; these don't exist in the model, so it's hard to get people who believe in it to recognise that they might exist in real life. Getting an economist to change their mind is difficult; the profession has a justified reputation for insularity and a lack of openness to outside ideas.

To an extent, economists are the way they are because they're cynics, and to an extent they're right to be; data that's not part of the model can be unreliable, dishonest or both. But even in situations where they have significant incentives to get it right, they still find it difficult to think outside the box.

None of these vices are ubiquitous; there are always exceptions. Economics maintains viability despite these structural issues simply by having a huge amount of resources dedicated to it, allowing it to generate huge amounts of surplus variety. Everything I've said about economics in this chapter has loads of counterexamples; there are so many economists that someone is always working on exactly the thing that you might claim 'economics' is ignoring. This means that its social and political position is unlikely to be challenged; it can always adapt, however slowly.

But this work tends to take place in universities, and there's a huge information-reducing filter between economics as it is practised in the academic world and the kind of economics that is used to address real-world questions. The amount of variety in modern economics is so great that nobody can be at the cutting edge in any more than a tiny part of it – even the most subtle theorist in one area tends to revert to the basic style of thinking in all others. Until an idea has reached the undergraduate curriculum, it isn't part of 'economics' for most useful senses of the word.

The real blind spot of modern economics is the economy.

Economists spent so much time on their general questions of optimisation and scarcity, price and quantity, that they seemed to forget that there's more to running a business, let alone a whole society, than just that.

Economists designed the world on one level, in terms of policy and governance; their mode of thinking became culturally hegemonic. But there's a whole world that economics hardly touched – and before we can diagnose the crisis of managerialism, we have to know a bit about managers.

7

If You're So Rich, Why Aren't You Smart?

To rule society, let it be remembered, is a full-time job.

James Burnham, *The Managerial Revolution*, 1941

Managers are the people who make the decisions that form the subject matter of economics; they're System 1 of a structure that has the economists providing some of its higher functions. That system needs to be able to transmit information between its levels, and it's likely to have bad consequences for higher-level stability if this doesn't happen. And this is indeed a problem.

There's a strange, broken communication channel between economics and management. We have people who study the production and exchange of goods as an abstract science, and there are people who work in organisations which actually make and trade things, but the links between the two are very weak. Some of the biggest problems of management are problems that economics not only fails to address, but seems to lack the tools to recognise. Many economists don't even speak the basic language of commerce – accounting – so they can't help manage the system, particularly when the accounting framework itself becomes a source of instability.

We don't do accounts

It always surprises people who haven't studied economics that most economists can't read a balance sheet. It sounds like the sort of thing that would be an elementary requirement, but accountancy is not part of many economics degree programmes. In fact, the higher the qualification someone has in economics, and the more prestigious the university, the more likely they've never cracked the spine on a set of company accounts.

The reasons are not dissimilar to the reasons why surgeons have separate professional bodies from physicians; there's a fear of being associated with a lower social class. Accountancy calls itself a profession, but there's a large element of routine work, and some aspects that come uncomfortably close to manual labour. One of my friends, as a young graduate trainee at a prestigious global firm of accountants, took part in the audit of a food company by putting on overalls and doing a stock count at a pig farm. If you're aiming to be considered a science alongside quantum physics, it doesn't do to be too closely associated with that sort of thing.

The first business school associated with a major university was founded with a grant from Joseph Wharton, in Pennsylvania in 1881 – he wanted to improve the dignity of managers by giving them a proper degree, rather than a certificate from one of the privately owned 'commercial colleges' that had sprung up across the United States. Lots of these colleges were quite grubby affairs, where manual labourers and shop clerks were taught the basics of bookkeeping and how to write a letter. Although modern business schools are much more prestigious and expensive, they're still regarded as 'vocational' rather than 'academic'; people worried about their status in the academy might very well see them as things to be kept at arm's length.

But it's not pure snobbery; there's also an intellectual self-defence mechanism. Accountancy, with its books of principles and standards, will always bring you face to face with the realities of production and distribution, and that sort of knowledge is difficult to reconcile with economists' modelling. Teaching young economists how to read a set of accounts might lead to awkward questions about time and information. Even an innocent question like, 'How do you really calculate depreciation?' could get in the way of an efficient modelling strategy that delivers a quick and definite answer. Best to protect the youth from such corruption.

Whatever the reason, economists don't do accounts – at least for the most part. And this matters a lot, because it creates a number of weaknesses in the higher functions of the overall system. We said in the last chapter that you might see the 'ist' in 'economist' as signifying ideology rather than study, but consideration of Stafford Beer's model ought to remind us how important ideology is in maintaining coherence. Nobody's really in charge of the global economy, but some of those who participate in it have the job of thinking about it in the abstract, balancing the unknown future with the present. It matters a lot if the people in charge of the overall structure, policy and governance are trained only in economics, and don't speak the same language as the management system that they're meant to be performing this function for.

Accounting is the language of management, and it has to negotiate a very particular ambiguity that Stafford Beer emphasised. This is the blurry line between System 2 and System 3, between the systems that make coordinated activity possible, and the integrating and optimising functions that determine what organisations do. The potential for confusion between these two concepts is intrinsic to the language used to describe them; words like 'administration' and 'management' cover both equally well. And it's an attractive ambiguity, not least because it gives you the ability to camouflage a decision as a necessity – which is a great accountability sink.

What are we talking about here? We're talking about numbers, considered as a means of control.

How Ricardians win arguments

The Ricardian Vice is a powerful thing. It is easy to fall into, hard to get out of, and works well for the people who have it. The way in which Ricardianism reproduces itself, and in which Ricardians win arguments, is that *they collect the numbers*.

Deciding on the methodology for data collection is a great way to make decisions without leaving fingerprints. Every decision about what to measure is implicitly a decision about what *not* to measure, effectively deciding what aspects of environmental variety are going to be ignored or attenuated.

More than that, if the data is going to be turned into information and gain a causal role in decision-making, it needs to be translated into the language spoken by the system. If you're in charge of designing and collecting the numbers, then not only do you decide what gets measured, you decide how it's classified and presented. So, for example, the death of a human being can be encoded as liver failure (medical classification), mercury poisoning (causal classification) or murder (criminological classification). Numbers are collected for a purpose, and it's often surprisingly hard to use them for any other purpose.

The thing that makes Ricardianism a vice is its tendency to prove things in a stylised model, and then act as if they were obvious truths about the real world. But if the model has been around for a while, the people who made the Ricardian

argument will have employed someone to collect data that val-
idates their model. Everything in the model will be backed up
with data; everything that the model leaves out will be 'difficult to quantify', soft and fuzzy. Unless you're lucky enough to
find data that's been collected for a different purpose, or unless
there's some fundamental inconsistency or flaw in the model,
you're very likely to look like someone who's trying to deny
cold hard facts – and that's how people lose arguments against
Ricardians.

It's easiest to see this happening in cases where the costs
are clear and immediate, but the benefits diffuse and longer
term. For example, you might think of an avant-garde theatre
company, which sells very few tickets and requires significant
subsidies to keep operating. On the numbers that go into an
arts funding body's spreadsheets, this could look like a bad
investment – a lot of public money is being spent to provide
small benefits. But part of the output of avant-garde theatre
companies is that they train directors of Hollywood blockbust-
ers. An overall flourishing arts scene has to include risk-taking
and difficult productions which only appeal to minority tastes.
And such a scene is a one of the vital commodity inputs to a
profitable media industry.

But this is very hard to prove, because the connection isn't
one that's had lots of work done on collecting numbers to
prove it. It's possible to put together case studies, and even to
do statistical work that suggests the link is there, but people
making the opposite case have a much easier argument – look
how much money is spent, and how empty the theatre is.

I described this problem in the context of economics, but
top executives in modern corporations are even more likely to
be susceptible to the Ricardian Vice. In a sense, a company's
accounts are a model; they're the financial control system in

written form. And if you can demonstrate something in the accounts, you're often a long way towards convincing people that it's true in reality.

This means that accounting makes a difference. And when there's a logical hole at the heart of the accounting system, that becomes an important blind spot. It helps our non-human decision-making systems develop a false view of the world. When that blind spot starts to interact with some of the other distortions created by the overall economic system, bad decisions start being made and feeding into each other. But before we get on to that, we need to understand the accounting in a bit more detail.

What are accounts for?

People collect numbers because they want to measure something. In the early days of manufacturing, numbers were collected by people who understood what they were measuring, and wanted to know how well it was being done. Railway operators wanted to calculate the cost per ton-mile, mill-owners wanted to know the cost per yard of a particular grade of fabric. This was the first purpose of accounts: the control and optimisation of a process. If you measure your inputs and your outputs, you'll know how efficiently you're transforming one into another. Over time, you can change things and see what effect they have.

Numbers like this are most useful when the physical efficiency of the process for turning inputs into outputs is the most important thing to be measured,* rather than the return

---

* Even then, you can screw up by measuring the wrong thing. A franchised chain of fast-food restaurants once judged its managers

on capital or the margin on sales. If, for example, you're a single-product company that sells its output to wholesalers at an easily observable competitive market price, and you're in a condition of constantly growing demand, so any investment will be paid back from ever-increasing revenues, then you can usually concentrate on physical efficiency and let the cash look after itself. The manufacturing and railroad companies of the USA in the late nineteenth and early twentieth centuries were in this position.

As time went on, though, companies got more complicated and investments got bigger. Companies started to make long-term wage contracts, to make multiple products and to take control of the entire production process, from raw materials to finished goods. The questions of 'overheads' and 'cost allocation' started to arise, inputs that don't directly or exclusively feed into a single process or a particular output – what happens to them in the system of accounts?

The cost of these inputs has to be covered, of course; prices might be set for each individual product according to specific market conditions, but the total revenue has to be greater than the total expense. And beyond mere survival, a company that makes multiple products needs to know which ones are profitable. That means examining those input costs, and deciding to what extent they're attributable to a particular output.

This kind of accounting is close to Stafford Beer's 'resource bargain'. In fact, it's the accountants' version of our

_____

on 'chicken efficiency', equating to the proportion of each chicken they were able to sell. Before long, the managers had worked it out – they stopped frying an hour before closing, telling customers they would have to wait for something to be cooked to order unless they took what was left. They lost lots of sales, but chicken efficiency was 100 per cent.

fundamental law of accountability; the extent to which a cost can be attributed to a product or output is exactly the extent to which the cost could be avoided if you weren't producing it. If you have a factory that makes nuts and bolts, for example, then the attributable cost of the bolts will definitely include the steel that they're made of. It will include the running costs of any specialist bolt-making machine and the wages of the bolt-makers. But if there are shared resources – machines that are used in making both nuts and bolts, say – then you have to share out the costs fairly. You could send someone to stand by the machine with a stopwatch and see how many minutes it's used in making nuts and how many in making bolts. But this might depend itself on the decisions you're making. If, for example, you produce large batches of nuts, but small specialised runs of bolts, then bolts are more costly, because they will take up more set-up time. If you change your production plan to make large batches of bolts too, the cost allocation will also change. At which point, there will, most likely, be a very big argument with the manager of the nuts division, if that involves allocating more of the machine's costs to them.

In the long term, all sorts of costs might be considered variable. Although the roof keeps the rain off every production process, reducing the product line might allow you to move to a smaller factory; there's an argument to allocate the cost of the factory rent proportionately to the space occupied.

But if the variability of a cost is dependent on the overall business strategy, the decision to allocate an overhead cost or not has to depend on whether the manager responsible for that product is really accountable for that part of the resource bargain. I would guess that a significant proportion of all stress-related diseases in managers are caused by having costs allocated to them that they can't do anything about. And some costs will never be allocated; you won't go far in your career if you ask the CEO to keep track of how many minutes she spends thinking about every different product, so as to precisely allocate her salary.

Process control and cost allocation are forms of what's known as 'management accounting'; they allocate expenses to either processes or products, and they're directly related to specific decisions. They're tools of variety amplification for the most part; the system allows an individual manager to monitor more processes than she could otherwise, but the information in the accounting reports is being provided in the context of an understanding of the processes. Or at least it ought to be; you have to learn about the numbers by looking at the business rather than vice versa, but a lot of people try to do it the wrong way round. This can be a real source of trouble, but it's a form of bad management practice that has grown inexorably over the last few decades, alongside the growth of another kind of accounting.

Accounting for growth

'Financial reporting' is quite different from management accounting, in a number of ways. And it's the dominant form of accounting today; when people hear the word 'accounting', usually in the context of an audit scandal, it's financial reporting that they usually think of. As the name suggests, it aims to report the performance of an entire company to the people who provide it with finance. From knowing that, we can guess that we shouldn't expect it to be consistent with either physical process measurement or cost allocation. The decision is what turns data into information; numbers collected for a different purpose are conveying different information.

Financial reporting also attenuates variety; it reduces and simplifies the process-specific information from the management accounts. In the first place, financial accounts have to be published to a regular schedule. The accounts refer to standard periods of time; these periods have to be the same for the whole enterprise rather than being matched to the control horizon of the individual decisions.

It gets worse, because there are two connected financial statements – an income statement which matches costs and revenues to the period of time, and a balance sheet which sets out the amount and value of the things that the company owns and owes. To complete the balance sheet, inventories of unsold and part-finished goods have to be recorded at their cost of production; as well as being matched to a period of time, costs have to be matched to units of output. A shocking amount of accountants' effort goes into this sort of thing; a survey of cost accounting textbooks once found that, on average, only a tenth of their page count was taken up with the actual questions of cost attribution, while 73 per cent was dedicated to the technical issue of inventory valuation.

But more than that, accounts need to be consistent, and this consistency is a form of information attenuation. The purpose of financial reporting is to *summarise* the performance of a business, not to reproduce it in every detail. Users of accounts need to be confident that they are produced on a consistent basis from year to year, and from one company to another. This means that most of the decisions regarding the allocation of costs are based on conventions – 'generally accepted accounting principles', a phrase which should immediately alert us to the presence of an accountability sink – rather than being matched to decisions.

For example, let's think about the question of overhead

costs and their allocation. Under generally accepted accounting practices, it's usually assumed that some categories of expense – marketing, debt interest and the CEO's salary – are 'period costs'. They're allocated to a period of time, and assumed to be spent at the moment the costly thing happens. Other costs, like direct labour and raw materials, are allocated to a unit of output. They hang around on the balance sheet for as long as that unit of output remains in inventory.

How might this go wrong? Consider a factory that produces skincare creams, with two main product lines. The first line is sold by the barrel to workers in the construction trade. The second line is basically the same stuff, but is sold for ten times the price in much smaller tubs to the beauty market.

Looking at a set of accounts drawn up using generally accepted accounting principles, the second product will look much more profitable on a per unit basis. Nearly all its cost base is in marketing, and advertising, which are usually considered period costs. It takes up much more space and time on the packaging line, but the company allocates this fixed cost in proportion to the direct labour cost, so the bulk product takes much more than its fair share. Whatever the cost of producing each product separately, it is a fairly sure bet that before long, an investor will ask why the company doesn't get out of the 'low-margin' business of selling to construction workers, and concentrate on the more profitable business of selling ten cents' worth of liquid paraffin for eighty dollars.

Einstein reads the phone book

This would only be a problem, of course, if companies allowed themselves to be fooled by the compromises made for their financial reporting, rather than keeping separate management accounts for themselves and using them for all decision-making purposes. Unfortunately, this often happens. You might think nobody would be so crazy as to make business decisions based on their effect on the financial reporting, but they do, all the time.

Basically, bookkeeping is expensive. It ought to be obvious to anyone who's had a job; the establishment of information and communication channels, and the translation mechanisms that support them, costs money. All the higher cerebral functions of an organisation involve spending resources on parts of the system that don't directly produce output and which aren't viable on their own. This was even more of a problem in the early days of the large corporation, before electronic computing became cheap.

Financial reporting is a legal obligation. If you're going to enjoy the many advantages of being a corporation, you have to produce audited accounts. But management accounting is an internal affair.* You can do what you like with respect to process measurement or cost allocation, but you *have* to employ people to collect numbers which are collated and systematised according to generally accepted accounting principles. In the days when this involved writing in ledgers and manipulating punch cards, the compromise was not too difficult to agree; the organisation would use financial reporting numbers, leaving divisional managers to do their own cost allocation and plant managers to collect process data for the outputs for which they were responsible.

In the 1960s, electronic computers became something which a large company might reasonably expect to buy and use. At

---

* You also have to keep a third set of books, for tax, but this makes the problem worse rather than better.

this point, a terrible decision happened – a decision which nobody made, but one with huge consequences. There might have been an opportunity to rethink the basis of financial reporting to adapt to such a potential change in information-processing capacity, but there was no incentive to do so; the user of accounts was still considered to be someone sitting in a chair and reading stacks of printed reports.

Companies might have considered the possibility of adding layers to their control systems, allowing process and cost allocation reports to be extracted from the same raw transaction data as the financial accounts. But that would have been a costly investment, and would have likely raised questions about organisation and hierarchy, as well as potentially overturning established relationships between cost centres and profit centres.

What actually happened was that most companies commissioned an electronic version of the same bookkeeping and reporting processes that they had been following with paper and punch cards. When Stafford Beer made his joke about recruiting all the greatest geniuses of humanity and setting them to work memorising the phone book, this was what he was referring to.

Business hallucinations

Mechanising and computerising the accounting system allowed companies to grow and get more complicated, but on the basis of the same system, one based on financial reporting. This meant that the increase in complexity was masked from the people who were meant to be managing it. Their reports were largely unchanged. And consequently, companies began to hallucinate.

Consider our imaginary skincare product company. We noted that the financial reporting numbers were likely to give the impression that the bulk product was much less profitable than the luxury one, whether or not this was true. An enterprising management consultant might ask why this company should make a product at all, if its value lies in marketing and branding. An interesting question to ask, and by no means necessarily one with an obvious answer. However, the way that the numbers are reported means that it's a question which is highly likely to get a particular answer.

Suppose that the skincare firm called in some management consultants to look at its manufacturing costs. Those consultants might start by looking at the cost of buying the bulk skin cream from an external supplier, rather than making it from scratch out of petrochemicals and vegetable oil. But that comparison is actually quite tricky to make. If the external supplier is mainly selling a commodity product to a small number of business customers, it's likely to have minimal marketing expense and only to do small amounts of research and development. Consequently, its overhead costs will be low. If the company's own manufacturing operations have overhead costs allocated to them in proportion to their direct labour costs (a pretty common method) or to their use of floorspace (another common method), then they will be taking a disproportionate share of the large overhead costs of the beauty business. And so, it's entirely possible that the decision will be taken to close down manufacturing, or sell it.

Of course, the marketing and research costs wouldn't actually have disappeared. In fact, the overhead costs might have gone up, as there is now a new set of problems relating to quality control and logistics which used to be handled as part of manufacturing. A couple of years later, the skincare company

might end up hiring another set of management consultants*
to explain why its star product seemed to have suffered such a
dip in profitability.

This seems laughable when you write it down in simple
terms; indeed, anyone who understood the numbers and the
cost allocation system would have killed the skin cream outsourcing project as soon as they became aware of it. But that's
easier said than done. For one thing, I've ignored the ambi-
guities of accounting in this hypothetical example. Even in a
very good management accounting system, the allocation of
overhead costs is *difficult*.

It's not just difficult in the sense that the spreadsheet is
hard to set up (although it certainly is that). It's fundamentally complicated because, as we said, whether a cost is fixed
or variable is a strategic decision; almost any cost is variable
if you're prepared to rearrange your business. Disagreements
about management accounting numbers represent disagreements about the actual production process.

This ambiguity tends to mean that the consultants in our
example have all the advantages of the Ricardian Vice on their
side. Their numbers are likely to be based on generally accepted
accounting principles. Anyone arguing against them would be
doing so on the basis of a customised set of figures of their own
design, with assumptions that are open to the reasonable suspicion that they have been cooked up to make a case against the
outsourcing transaction by a change-resistant middle manager.

The problem here is that unless a lot of effort is expended,

---

* Who am I kidding here? It would almost certainly be the same
consultants, coming up with a different answer. Generating repeat
business is a key skill for a partner at a consultancy firm, and it
wouldn't be difficult to come up with a plausible explanation that
didn't involve admitting anything about the last project.

it's easy to create an information system that will always give a particular answer, whatever the truth is. And that answer will appear to be an objective fact, even though it's actually the result of a lot of implicit assumptions. Once more, we see that important actions can end up being the consequence of decisions nobody made – or, even worse, decisions that people made without realising they were doing so. An accounting system is an almost perfect accountability sink – even the people responsible for constructing it don't necessarily understand what they're doing.

Structure, strategy and struggle

Given all this, it might be considered surprising that large companies don't end up in a worse state. But, of course, most companies, quite fortunately, don't make every decision based purely on numbers. They have managers, who provide the necessary variety and context to mitigate the inevitable distortions in the accounting system. Often, a core skill of middle management is the ability to manipulate the financial reports to compensate for a set of numbers that aren't giving the right answer. Everyone who has put together a business plan knows that if you can't fudge the key assumptions to justify the decision your boss wants to make, you don't know enough about the business.

Of course, this is not in itself unproblematic. In this slightly parodic picture of a big decision-making system, we have a compromised set of communication channels between black boxes, which is compensated for by adding resources in the shape of middle managers, who provide the capacity to translate the skewed signals from the financial reporting system into more or less accurate and actionable information.

There are two obvious ways to fail here. The translation and error-correction mechanism might be inadequate, or the managers might intentionally distort the signals in order to follow priorities of their own. The tragedy of senior management is that it can drift into either of these failure modes without realising; if either problem arises, it arises in their information and communication environment, so they won't notice it. It's the problem identified by Niccolò Machiavelli – a prince who is not wise cannot be well advised, and a manager who doesn't have access to excess analytical capacity won't be able to tell when something has gone wrong with their subordinates. But maintaining that spare management capacity is expensive.

Every new idea in management seems to come as a reaction to the previous generation's attempts to find the right way to solve these two problems without taking on too much in the way of overhead costs. And for the same reason, the management practices of previous decades will always seem hilariously dated. The complexity of the environment is always increasing, and technological progress means that there will always be new ways of getting information from where it is generated to where decisions are taken. *Of course* management theories change – when the foundations of the problem are constantly shifting, the answers are bound to change.

If you translate this problem into the more abstract language of cybernetics, it becomes easier to understand. In the business environment, complexity (environmental variety) will naturally increase. In any given organisational structure, the variety which management systems can bring to bear also increases, but more slowly.* At some point, the difference in

---

* If it was possible to measure variety in absolute terms for real-world

growth rates becomes critical; the organisation needs to change its architecture so that variety is matched to variety at each level of decision-making. John Paul Getty of Getty Oil was able to read status updates from all of his rigs every morning in the 1930s, but that's simply not possible for a company like BP or ExxonMobil today.

If you read the great books of management with this in mind, you'll notice a strong common theme; the stars of this literature all try to get managers to understand that they must create systems which regulate themselves rather than requiring constant supervision. Most of what's worth reading in management science is about stopping decision-making systems from becoming overwhelmed.

Alfred D. Chandler, a historian and management scientist, saw this. In *Structure and Strategy*, his book about the development of American companies including DuPont and General Motors, he realised that as they became more complex, they had to develop a multidivisional structure. Almost as soon as his book was published, though, this structure was itself reaching the limits of its capacity. And this story repeats itself through the history of management science; almost every classic of the literature seems to have described a way of adapting systems to a more complicated world, and then to have become obsolete itself. If you look past the slogans and think about what things

---

applications, there might be some hope of a sort of 'law of motion'; one might find that environmental variety grew exponentially and management variety only linearly, for example. But this is a dead end for really big systems; the kind of quantification we'd need is not possible because everything is already at the level of 'unthinkably huge'. All we know is that in the end, the environmental variety must have grown faster, because the system became increasingly unregulated.

like 'management by objectives', 'focus on core competences' and so on actually mean, they are all different ways of advising executives to restructure their businesses so that they don't generate complexity faster than it can be managed. Meanwhile, the management consultancy industry prospered by selling the myth that it might be possible to economise on management capacity by renting it when you needed some rather than paying it a salary.

The guys you love to hate

Management consultancy is not a popular industry. Even in companies that spend a lot of money on its services, the typical view is cynical – people say that it means borrowing the client's watch to tell them the time, for example. But when you think about the services management consultants provide, you quickly realise that being the butt of that sort of joke is intrinsic to the nature of the work.

In the simplest sense, it's intrinsic because management consultants often act as accountability sinks. It's not uncommon to be in a situation where the management of a company need to do something, but know that if they do it, they will damage their relationship with the employees so badly that they won't be able to manage any more. In that situation, you need a professional scapegoat – you could hire an out-of-work actor to read the solution from cue cards, but management consultants are better at pitching for the business.

Even if they're not acting as scapegoats, though, management consultants will often work by telling a company something that its employees already know. Stereotypically, a consulting assignment goes through stages of signing the contract, talking to the employees and middle managers, finding the person who knows how to solve the problem, then packaging up their solution and selling it to top management. In one sense, the consultant isn't contributing anything new. But in that case, why wasn't the solution already being implemented?

We can see, having considered cybernetics, that there was clearly a feedback loop that wasn't working – either the communication channel wasn't there, or it didn't have the translation capacity, or there wasn't a working intelligence system at the other end. All these things are management problems! So if a consultant carries out the assignment as described, they have solved the management problem and done what they were paid for. In this sort of situation, criticising the consultant for selling the company's own insights back to them is like complaining to your hairdresser that they've only cut away something that was yours.

When things go wrong with management consultancy, it's more likely to be because the consultants are tackling a new problem and there isn't anyone in the company who knows the answer. The consultants get commissioned because they advertise themselves as brains for hire; a company can cut overhead costs by having fewer middle managers performing staff functions in the ordinary course of business and buying in brain power when confronted with a difficult question. When you write this idea down in black and white, it's pretty easy to see why it won't work except by pure luck.

The reason is invariably a failure to respect the complexity of the problem. Management problems are complex, high-variety questions, Rubik's Cubes rather than rows of blocks. In order to solve them, you need to make decisions about how to represent the problem in such a way that you can simplify it and solve it, without losing vital details that will blow your solution apart as soon as it's implemented. If the systems aren't in place, understanding the problem well enough to create them will require a large initial investment; you're having to reproduce all the expertise of the middle managers you thought you were saving money by not employing.

In the context of a consulting assignment, this investment will translate into billable hours – which accounts for management consultancy's reputation for being absurdly expensive. There's really no way around it; you either end up employing about as many people as you would have done anyway, but on consultancy firm hourly rates rather than middle manager salaries, or you end up accepting generic strategies based on what a consultancy firm partner has learned elsewhere.

The generic solutions are not always bad. A lot of management problems *are* surprisingly generic: a missing communication channel or a failure to notice that the environment has changed. If the problem is understood correctly, the solution is usually quite simple – it's likely to be a variation on the same theme of 'the world has got more complicated and you need to take steps to reconcile that complexity with your capacity to manage it' that has been the message of management science since Alfred D. Chandler.

But because the solutions are often simple, the work is surprisingly unpleasant. An effective consultant is likely to spend most of their time telling people obvious things that they don't want to hear. That's a difficult combination; while not particularly intellectually stimulating, it's emotionally taxing. It's not surprising that so many people find doing this intolerable, and consequently let their ethics slip. Telling your client what they want to hear is a better way to get repeat business; the problem won't go away and the person commissioning the work will still like you.

Ideology of the managerial class

This is why the global management consultancy profession can't provide corporations with the identity-creating function that economics provides to policymakers. It spends too much of its own capability and information-processing capacity in telling people what they want to hear.

Economists do this too, of course, but they are able to call on a huge reservoir of brain power in the universities. Business schools, on the other hand, have much more emphasis on teaching than on research. Only one or two professors in every generation can come up with a new way of rephrasing the generic problem of management, the fact that organisations have continued to get more complicated at a faster rate than the managers have. If you're not lucky enough to be in that small number, you either have to become a disciple of one of the big stars, a specialist in a small niche, or to turn inwards and write for an audience of other academics.

Consequently, a large proportion of academic management science seems to consist of articles and books bemoaning the lack of a coherent intellectual framework or a connection between theory and practice. As a discipline, it has a hard time deciding whether it's meant to be a critical social science *about* management, or a practical one *for* management. Many of its top practitioners have an eye on the lucrative market for books that sell well in airports,* which makes them reluctant to participate in research programmes that can't be incorporated into their personal brand. And when a big idea comes along, it tends to get overused and applied in inappropriate contexts.†

---

* Or at least, used to; in-flight WiFi and laptops seem to have more or less consigned this once useful cliché to history.
† Management science also has a fairly big problem of what might

In many ways, this is a natural state for a social science to be in. There's no single coherent intellectual framework behind sociology or anthropology either, and the cutting edge of academic thought in political science has as little influence on politics as management science does on business. Economics is the unusual case, in its ability to provide a framework and ideology for an important part of the global decision-making system.

The failure engine

To summarise: if we think about the main decision-making systems of the industrialised world in cybernetic terms, we can see that there are a number of important blind spots and dysfunctional systems. In economics, we had the Ricardian Vice, and its consequent problems with time and uncertainty. Outside economics, in the wider world of management, we have another version of the Ricardian Vice, based on the misuse of accounting systems. We also have a problem that managers are structurally overwhelmed by complexity, and they don't handle it well because they lack a coherent ideology and sense of themselves as managers. Of the two groups that might have been capable of providing this function, academic management scientists are just too weak and divided, while management consultants are too caught up in telling people what they want to hear.

These problems feed into each other – the blind spots of

---

delicately be called 'replicability problems' and less delicately 'pervasive research fraud'. This is not a modern phenomenon; it's pretty well established today that the creator of the field, Frederick Winslow Taylor, faked nearly all of his numbers.

economics and the blind spots of management work together to produce a model of the world that can drift away from reality, and start producing bad decisions. There's no self-correcting tendency, because the problem is in the information-processing system itself. All the incentives and failsafes – what Stafford Beer would call the 'homeostatic forces' – are operating to preserve the system as it is, not to adapt it to the facts. Consequently, as I've begun to suggest, the non-human decision-making systems of the world have begun to go a little bit mad.

INTERMISSION

Meanwhile, in Chile

The short-lived government of Salvador Allende, and the Pino-chet coup which ended it, were both parts of a huge event in history. It was a pivotal moment in cybernetics, and one of the defining episodes of Stafford Beer's life, but in the grand scheme – the Cold War, Latin America and the trauma inflicted on the people of Chile – the cybernetic project was a footnote and so was Beer. I'm not a historian, I'm not Chilean, and I'm a fundamentally quite whimsical person. It's really not for me to write the story of Pinochet and Allende.

All I'll say is that, from a cybernetic point of view, violence is a horribly effective variety amplifier. A single guard on a watch-tower can't do much; give them a gun and they can control an entire prison. If you have a system of gulags, you don't need to meet the arguments of your opponents one by one. A state which is not supported by a public consensus about its legit-imacy, and which is consequently overloaded by its political environment, is highly likely to see atrocity as the only means available to maintain its stability.

The story of Stafford Beer in Chile, though, is worth understanding because it shaped the world's understanding of cybernetics and the reception of his theories, and it's generally quite badly understood.* The great misconception about the Chilean experiment is that it was an attempt to apply computer technology to the task of organising a command economy. This is a handy myth for those who want to use it as an example for whatever argument they're trying to make. Neoliberals and apologists for Pinochet and Friedman can pretend that it was a quixotic project with no more hope of working than the Soviet system; socialists and other apologists can pretend that it nearly worked. Neither is the case, and by now it ought to be obvious why – management cybernetics demonstrates quite clearly why a centrally planned economy doesn't respect the law of requisite variety and can't get the information it needs to function.

Instead, Beer helped to set up what amounted to a massive system of 'management by exceptions'. The overall project was called Cybersyn (cybernetics and synergy), and it was meant to help the government manage their newly acquired nationalised industries, without depending on their previous owners. It had three main components – Cybernet, Cyberstride and the Operations Room.

The Operations Room was an ergonomically designed command centre; pictures of it look very much like the bridge of the USS *Enterprise* in *Star Trek*, apart from the ashtrays and drinks holders. It was meant to be occupied by about a dozen managers, who were able to call up data and charts on the big display screens that covered the walls.† On the basis of highly

---

* The publication of Eden Medina's *Cybernetic Revolutionaries*, a very good history of the project, has helped in recent years.
† These were not computer display screens – there was a team in the background of the Operations Room who would hand-draw charts on slides to project on to them. Stafford Beer claimed that this was an advantage rather than a compromise; he thought that the 'flicker' of cathode ray tubes was distracting.

focused discussion and debate, they would send instructions to the management of the factories.

But they would not debate every single production instruction for every business in the Cybersyn network, of course. Selecting the subjects for the Operations Room to think about was the purpose of Cyberstride. This was a piece of computer software, which took in data and used statistical analysis to see whether it was in the normal range of variability. It could then predict whether any departures represented a temporary blip, a change in trend or a new level of normality, and escalate them accordingly. Effectively, it provided an automated and expanded version of the same function as the trellis of hand-drawn charts which had first built Beer's reputation in the steelworks – it identified potential areas of concern.

The input to Cyberstride came via Cybernet, which was a network of hundreds of telex machines. These had been fortuitously discovered lying idle in a government building, but they were perfect for the task of transmitting production data – much more reliable than reading numbers out over the phone, and generating an automatic record of the past. It turned out that this communication network was the most practically useful part of the system; during a trucking strike, it allowed the government to monitor deliveries and ensure essential supplies.

If you look at Cybersyn along the lines of the viable system model, you can see that the control room is playing the role of System 3 (negotiating and renegotiating the bargains with the operations), with Cybernet having some System 2 roles in preventing clashes and coordinating resources. Cyberstride is being used as a piece of variety engineering to help match the resources of the Operations Room to the data coming in. There was also a plan to develop another piece of software (Checo, the Chilean economic simulator) to act as a System 4 and allow potential future and outside shocks to form part of the debate in the Operations Room. We'll talk about System 5 later.

But, if you look at Cybersyn in terms of management cybernetics, you'll immediately suspect that the whole project was badly out of scale to the problems it was meant to solve. While the exact measurement of information and variety is almost impossible at large scale, the idea that a few hundred telex machines might be matched to an industrial economy just doesn't pass a laugh test. They did extremely well to manage the trucking strike, but as a test of the viable system model, this was set up to fail.

Furthermore, the very act of creating Cybersyn provoked a reaction from the system it was meant to regulate. Even in the early stages when Beer was first retained as a consultant, people had been warning the Allende government that it was nationalising things too fast, without regard for the capacity to manage them, and seemingly without much in the way of an overall industrial plan. They had introduced a system of 'interventors' who sat in the factories and passed on orders from central government; this was unpopular, and contributed to the impression of central technocratic control rather than devolved autonomous production. The biggest difficulty that Beer and the project team faced was getting cooperation from the field.

And the Cold War itself cannot be wholly ignored. Chile was the subject of de facto economic sanctions from the USA (including restrictions on the export of computers and tele-communications gear). Hard currency reserves were scarce, and made scarcer by the Chilean capitalist class's fear of Communism, which led them to move money offshore and liquidate industrial investments. The Cybersyn project didn't credibly have enough bandwidth even to run the economy in a steady state – it certainly wouldn't have been able to cope with the kind of rolling crisis that Chile actually faced in the early 1970s.

But then, it turned out, that all these questions would go unanswered, because the Pinochet coup happened.

This had a huge effect on Beer; we'll come back to his post-Chile career later. He spent years trying to help the members of the Cybersyn team escape to physical safety, and his writings took a turn for the dramatic; rather than textbooks of management science, he started addressing fundamental issues of democracy and society. At the age of fifty, with most of his children grown up, he left his business and retired to a small cottage in Mid Wales.*

I said earlier that more lessons should have been learned from the failure of Soviet cybernetics. In the case of Cybersyn, it didn't even get the chance to fail in a way that could have taught us anything.

---

* It was called Cwarel Isaf. It's been restored today, by a foundation dedicated to preserving Stafford's ideas, which in turn is sponsored by a management consultancy firm. Apparently, you can stay there for a few weeks if you're doing research into cybernetics.

PART FOUR

WHAT HAPPENED NEXT?

8

Enter Friedman

Only a crisis – actual or perceived – produces real change. When that crisis occurs, the actions that are taken depend on the ideas that are lying around.
  Milton Friedman, *Capitalism and Freedom* (3rd edn), 2002

There's a huge temptation to turn Milton Friedman into a monster, but he wasn't one. He was a university professor and Nobel laureate, the intellectual guru to a generation of free market economists, that's all. If he hadn't written the essay 'A Friedman Doctrine: The Social Responsibility of Business Is to Increase Its Profits' in 1970, it is likely that someone else would have. But it *was* he who wrote it, along with *Capitalism and Freedom*, and dozens of newspaper columns. It was he who was the academic sponsor and mentor of the 'Chicago Boys', the economic advisors to the Pinochet government. Other people also brought libertarian ideas into the mainstream of American political culture, but none of them were as famous. Friedman was about as perfect an example as one might get of an economist who worked as part of the highest-level governance system of the capitalist world.

The word 'neoliberalism' is atrociously contested in terms of its meaning; something that so many people believe in will always be impossible to boil down to a core definition of the sort you could print on a T-shirt. But in so far as there is one, neoliberalism was a matter of shareholder primacy in corporate governance, combined with deregulation, privatisation and low taxes in public policy. And there's no way to tell a story about those economic concepts without Milton Friedman being the main character.

I've discovered over the years that lots of people liked Milton Friedman and many more admired him intellectually; as a result, discussions of the harmful effects of his legacy tend to turn pretty angry, pretty quickly. The important thing is to remember that the extent to which Friedman's ideas were replicated and enacted by the global political and economic system has more to do with the system than with him. They might have been inspired by him, but they were still for the most part decisions that nobody made.

The thread running through the ideological shifts associated with Friedmanism – the neoliberal revolution in public policy and the post-1970 turn in managerialism – is that of a deep, almost nihilistic lack of trust in human judgement. If fallible people were put in charge of something, they would generally be lazy, careless or dishonest; the only way to make it otherwise was to harness their selfishness, making sure that they were personally bearing the costs and gaining the benefits. The market transaction was important to Friedman because it was the only human relationship that you could be sure was honest. A famous quotation of his sums it up:

> There are four ways in which you can spend money. You can spend your own money on yourself. When you do that, why then you really watch what you're doing, and you try to get the most for your money. Then you can spend your own money on somebody else. For example,

I buy a birthday present for someone. Well, then I'm not so careful about the content of the present, but I'm very careful about the cost. Then, I can spend somebody else's money on myself. And if I spend somebody else's money on myself, then I'm sure going to have a good lunch! Finally, I can spend somebody else's money on somebody else. And if I spend somebody else's money on somebody else, I'm not concerned about how much it is, and I'm not concerned about what I get. And that's government. And that's close to 40 per cent of our national income.*

To reiterate; this isn't just Milton Friedman; this is the system talking. Friedman is important because he was so intelligent and quick-witted that he summarises a whole complex of ideas here, in a form that was easy to pass through the communication networks, the better to build identity and shape policy. The message was wonderfully crafted for this purpose. But it was this message that took off, because Friedmanism in this sense was a solution to a number of incipient crises in a high-level decision-making system – a very slow, very large artificial intelligence consisting of and regulating the Western industrial economies. The capitalist world had reached a stage where it needed Milton Friedman.

Hippies, considered as cybernetic disturbance

In the early days of the post-war corporation, the question of purpose hardly arose. Companies like DuPont had been

---

* This version of the quote comes from an interview with Fox News in 2004, but he said it several times, and a less pithy version of the same idea is in his book *Free to Choose*.

given an obvious reason to invest in the production of smokeless gunpowder by the Second World War, and at its end they found themselves in possession of a lot of capital equipment and chemical know-how. They went out looking for new things to do with cellulose and petrochemicals because it would have been strange not to. As they grew more complicated, they had to reorganise their corporate forms and management structures; academic writing on management was really just catching up with the things that engineers and accountants were inventing out of necessity.

As the 1950s turned into the 1960s, writers like Peter Drucker started to look at the role of the corporation in society. And this wasn't just the purview of business thinkers either; the development of the 'Organisation Man' and the rise of the professional and managerial class as an entity was a big thing, the novelty of which is hard to understand today. Novels like Sloan Wilson's *The Man in the Gray Flannel Suit* were bestsellers. People really regarded the new life pattern of university, office work and suburban living as something new. They also started to worry about the ways in which decision-making was changing – the word 'groupthink' was coined in 1952.

One consequence of the post-war baby boom was a corresponding explosion in the variety of ideas and priorities in the economic and political environment when those children reached adulthood. Throughout the Sixties, people began to notice that the climate was changing. Theodore Roszak coined the word 'counterculture' in 1969, but his book *The Making of a Counter Culture* is remarkable for the extent to which it's based on contemporary magazine articles, news reports and panel discussions. Everyone knew that something was going on. American business felt that it had to adapt to the values of the new generation.

However, there was no mailing address to write to if you

wanted to find out what those values were; the defining characteristic of youth culture was its variety. Trying to navigate and
understand how the world was changing was an almost impossible task. In order to restore balance, the system needed to find
some way of reducing the variety. It needed to find out who to
listen to, yes, but it also needed to be given guidance on who to
ignore. Eventually, the nation turned to Milton Friedman.

J. K. Galbraith, possibly the only other economist of the
1970s with anything like Milton Friedman's public profile, once
said that the essence of leadership was 'the willingness to con-
front unequivocally the major anxiety of their people in their
time'. If we regard Friedman's 'people' as corporations, their
major anxiety at the end of the 1960s was that they lacked a
sense of purpose. On 13 September 1970, though, the answer
landed on America's doorsteps.

It wasn't exactly a new argument on the face of it, and nor
was it unique. But Milton Friedman's 3,000-word essay on the
question of social responsibility in business set out the case
so concisely and so attractively that it shaped the culture of
management and business for decades to come. From then on,
the question of the purpose of the corporation was more or
less settled. Essays and panel sessions on 'ethical capitalism',
'corporate responsibility' and 'stakeholders' would continue
among successful people with socialist or Christian parents. But
although we tried to take the opposite case seriously, everyone
really knew that corporations serve the purpose of shareholder
value, and that was an end to it.

The Friedman doctrine

If you want to understand how things began to drift off course
during the neoliberal era, Friedman's essay is worth studying in

detail;* it's a masterpiece of persuasive writing and thoroughly deserving of the fifty-year retrospective published in the *New York Times* in 2020. What's interesting, though, is that most of the specifically political and philosophical content comes surprisingly late in the essay, almost as an afterthought.

The really persuasive case is made much earlier, as an address to the self-interest and self-image of 'corporate executives'. In the middle of a peroration about the difficulty of the socialist calculation problem and the impossibility of knowing the true 'social good', Friedman reminds executives that they can be fired if they don't put profits first. Although it's often supposed that the doctrine of shareholder value maximisation has its roots in libertarian philosophy, the Friedman doctrine seems to put things the other way round – libertarianism gains credibility from its association with the idea that corporate managers' true responsibility is to increase their profits.

The argument is effectively completed in the essay's first ten paragraphs, in which Friedman sets out an entirely different view from the management cyberneticians – one where the organisation doesn't really exist at all. Friedman describes the company as an 'artificial person', but this is only for the purpose of describing its responsibilities as 'artificial responsibilities' and making the point that if the company's responsibilities are artificial, then 'the business community' as a whole certainly can't have any. In the Friedman doctrine† the

---

* How much detail? Well, this section is about 1,500 words. On the face of it, that's about a 50 per cent variety attenuation because it's about half the length of the original essay. In all honesty, I'm saving you less time than that; a significant proportion of Friedman's essay consists of arguments against forgotten contemporaries which you'd be able to skip anyway.

† Although the phrase 'Friedman doctrine' has been applied over the

only possible locus of responsibility is the individual corporate manager.

So the protagonist of 'A Friedman Doctrine' is the same as the hero of Burnham's *Managerial Revolution* – that large professional and managerial class which grew rapidly during the twentieth century and which is distinguished by the fact that it exercises control over capital assets without necessarily owning them.

Equivalents to these people were there in Karl Marx's day – they show up every now and then in *Capital* as 'supervisory labour' – but they were never fully analysed. Interestingly, though, Friedman agrees with Marx on the class position of the executives – they are proletarians, whose relationship to the owners of capital is simply that of employment. He just thinks that's a good thing.

This is the whole theory of the firm, according to the Friedman doctrine. There is no analysis of the company as a decision-making system, just individuals making decisions. While modern HR people might talk about 'bringing your whole self to work', Milton Friedman explicitly tells the (assumed to be male) modern executive that he has to divide his moral self into judgements 'as a person in his own right' and decisions 'in his capacity as a businessman'.

The justification for doing this is quite surprising. One might have expected Friedman to portray the socially conscious executive as a pickpocket, stealing the marginal profits which the shareholders might have relied on for their dividends, but that's not the line he takes at all. If you want to speak to the

---

years to almost everything he ever wrote about, from monetary policy to privatisation, the only example I can find of Milton Friedman using the specific words himself is in the title of this essay.

deep fears in the soul of modern management, you can't throw around vulgar abuse; you need to force managers to face the thing that they're scared of, and raise the charge that their fear might be valid. The greatest fear of the manager is, of course, management.

And so the crux of Friedman's doctrine is that when companies act in the interests of society instead of their shareholders, they take on the role of a *government*. The benefits that are conferred when a boss fails to make sufficiently aggressive redundancies, keeps the price of life-saving drugs affordable or allows employees a day off to vote in an election – this is the moral equivalent of *taxation*. If the appointed agents of the owners of capital do anything other than promote the interests of their principals, they are taking the decision to spend other people's money, entering into the 'political mechanism' of control and coercion rather than the 'market mechanism' of free and voluntary exchange.

In other words, the Friedman doctrine is a doctrine of the claims of capitalism against managerialism. In the essay, he argues that there is no such thing as 'society', only voluntary associations of sovereign individuals. But what he doesn't say is that the argument requires that there are no such things as corporations either, that organisations have no independent status and should be looked through as if all their staff and management were in a direct relationship of employment with the business owners.

I've always disliked this doctrine; it seems to me to be a sneaky way for shareholders to try to get the benefits of legal personhood and limited liability without giving anything back in return. If they wanted to be nothing more than a voluntary association of individuals coming together to buy some factories and hire people to run them, there are plenty of ways to do that. If you pick a form of organisation that has legal privileges, you can't just assert that the officers of your company still owe you all the same duties as if you'd personally hired them; the legal and political compromises which created the limited liability corporation were made on the basis of a different set of expectations about the behaviour of corporate citizens.

And looking at the Friedman doctrine from the cybernetic point of view, there's another dimension that makes it utterly unsustainable: he's asking corporate managers to pretend to do something that they must know is impossible. To think of the corporation as a set of relationships between individuals is absurd – there are far too many of them. If executives were really to see themselves as the personal agents of individual shareholders, they'd have to consider a huge number of possible relationships. A large corporation could have thousands of shareholders. Those shareholders could include people with widely different interests, particularly since many of them will also own shares in competing firms. Many companies are at least partly owned by their workers' pension funds. And the shares trade every day; there isn't even a stable group of individuals.

The Friedman doctrine forestalls this cacophony of complexity by substituting an abstract 'representative shareholder'. Rather than thinking about actual human beings working in association, we're invited to replace them with a black box that only cares about profits, then pretend that we can enter into the same sorts of relationships with that black box as we could with the individuals.

This is what makes the Friedman doctrine, as far as I can see, a lie; it's an exhortation to executives to reverse the truth. They are meant to ignore the reality of the company and act as if they are directly employed by human beings, but then ignore the reality of human beings, and act as if they are employed by a theoretical construct.

It's an attractive lie, though, thanks to the combination of the accountability sink with the shift in perspective. The great anxiety of the managerial class was that they were losing their individuality as corporations became more complex, but that they were still subject to criticism. The Friedman doctrine invited them to disassociate themselves from their roles; to attribute all the bad consequences and all the frustrating lack of independence to a separate work-self, which was under an obligation to a simple principle.

There is a lot of quite dark and sophistical stuff in the essay after this; having described the potential for socially responsible business as a 'suicidal impulse', Friedman sets up a slippery slope and starts arguing that if you tolerate a little bit of do-gooding by corporations, you'll soon 'strengthen the already too prevalent view that the pursuit of profits is wicked' and thence onward to Communism. But the key move was the first one made, and it's no wonder that it caught on. It was so terribly attractive to, it solved so many problems for and soothed the anxieties of, so many important people.

The doctrine in the world

Over time, the Friedman doctrine was refined in odd ways. Rather than a mock 'social responsibility' or a real employment contract, people started talking about a 'fiduciary duty' of management to 'maximise shareholder value'. This makes no sense; it's legal jargon being used outside its specific legal context.

Directors and officers of companies do have some fiduciary duties – they have to act in good faith with the assets they've been put in charge of, they can't self-deal or moonlight for another company and so on. But the courts have never found that there's any particular fiduciary duty to run a company in one way rather than another, and there's certainly not one to maximise the value of one particular class of its financial securities. It would be a legal disaster if they did; what if the way to maximise the share price involved taking a huge risk of bankruptcy and ripping off the creditors?

When the 'fiduciary duty' formulation was first used is unknown. It probably got picked up from legal judgments on actual cases of fiduciary duty (mainly in takeovers, where the board does have a duty to get the best deal possible) by people who thought that it sounded like a clever and impressive way to emphatically say 'this really, really, really is a duty'.* People today quote the modern version as attributable to Friedman and even cite this phrase to his essay, without checking what he wrote. When something like this happens, it's pretty easy to guess that ideology is at work.

Of course, the purpose of the phrase 'fiduciary duty to maximise shareholder value' has never been to make a clear claim about the law; the language is there to dress up a moral exhortation in sonorous tones, to make it sound like an ancient obligation that's blasphemous to question. The legal language is meant to suggest to executives that they're part of a system which demands that they separate their decisions as individuals from those as employees of the shareholders. The economic

---

* This happens a lot. Compare, for example the 'duty of care', regularly used to insinuate something like a parental relationship where a company is meant to be watching over you. In fact, it's short for 'duty to take care'; the canonical case is one in Scottish law where it was found that a ginger beer manufacturer had a duty of care to ensure that small snails didn't find their way into the bottles.

language is there to focus all the different priorities that they might have on a single maximising end. The executives get the benefit of a clear conscience; the Friedman doctrine is also an accountability sink for them at the personal level.

It wasn't just the relationship between managers and shareholders that experienced semantic drift. The word 'increase' was replaced with 'maximise', to make the language consistent with that of economics,* and to distinguish the new value system from the old days of DuPont and General Motors when managers sought to earn a reasonable return on capital and grow the business every year, but didn't necessarily cut costs aggressively to manage the business cycle. And rather than 'profits', managers were directed to maximise 'shareholder value'.

The shift from profit to value was partly a solution to a problem and partly a reflection of how the economy was changing. As we saw in the last chapter, 'profits' are not a simple thing. When a slogan like the Friedman doctrine is passed on from economists to managers, tricky questions arise. 'Maximise profits in what sense?' 'According to what accounting standard?' 'Over what period?' 'With what sort of trade-off between present return and future growth?' 'What kinds of risks should we be prepared to take?'

In large companies, rather than getting involved in the answers to these questions, managers could be pointed in the direction of their share price, a real-time computing machine that delivered a positive or negative verdict on their strategy.

---

* This one is easier to track down to a moment in time – Michael C. Jensen and William H. Meckling's 'Theory of the Firm: Managerial Behaviour, Agency Costs and Ownership Structure', in the *Journal of Financial Economics*, 1976.

Friedman and financialisation

If they happened to work for a company that wasn't big enough to have its shares quoted on a stock market, 'shareholder value' was a more abstract concept. Managers might construct spreadsheets to imagine what might have been happening to their share price, but this wasn't really satisfactory; it was too much like marking their own homework. There are other ways for finance to act as a means of discipline, though, and the post-Friedman economy made very great use of them.

We haven't talked much yet about the fact that the financial sector has grown massively in importance during the period we're interested in, the time in which things started to get out of control. But it's always implicitly there; it's part of the overall system of governance and regulation of the modern industrial economy. The way 'the financial system' works is through nearly all economic activity having some financial aspect to it. Finance is how the resource bargains are struck, it's how commitment is made to plans, it's a key way in which information is transferred.

This makes 'financialisation' a bit of a misnomer. It's not the growth of a parasitical separate system; it's the increase in importance of the financial aspects of the overall economic system, as it grows in size and complexity. Banks and financial institutions only really exist as sets of relationships with their clients. The fact that you can create a company from a set of these relationships and put its logo on big office buildings is one of the most important ways in which traditional organisation structures can misrepresent the underlying transfers of control and information that govern the arrangements of production.

However, because finance is involved in all the levels of recursion, there are financial aspects at the highest levels of control and management. In particular, the way that the market economy manages the highest-level decision of them all, the balancing of present against future – that's handled in the financial markets. When companies sell shares, or issue bonds or borrow money, they are reaching out to that market to get a decision from the economy as a whole about how and on what terms it wants to allocate current resources for investment in future production.

And financial markets do give answers to these questions, thousands of times a day, in a way that can seem miraculous. When you look at the capital market in this way, you might be briefly overwhelmed, to the extent that you wonder whether it might be a magical computing fabric after all. It's an extraordinary thing; walking across a trading floor is a tangible experience of what it might be like to live inside an artificial intelligence. This sense of getting out more than you put in certainly helped the Friedman doctrine to gain acceptance; as well as being a wonderful accountability sink, 'the markets' do feel like they might be better decision makers than any individual human being, or any small group of people that you might be able to gather together in a boardroom.

Debt as a control technology

If the market is the brain of the capitalist system, its nerves and muscles are made out of the debt relationship. Starting in the 1970s, the economies of the industrial world saw the beginnings of a huge increase in the use of debt, and this might have been the most significant decision which nobody ever made. Debt may have ended up as a problem, but it always starts out as a solution.

Debt is often poorly analysed by economists, largely because the dominant method of analysis in economics is through equilibriums and, in equilibrium, not much happens with debt

– money gets paid one way, and then paid back the other way later. Interesting things happen with debt in situations where expectations have *not* been satisfied, one way or another, and that's something that's quite hard to talk about if perfect information is your starting point. We said back in Chapter 6 that economists deal with time and uncertainty by flattening everything out into sets of expectations over counterfactual possibilities – but debt and bankruptcy don't have well-behaved probability distributions. Corporate failures don't really form a homogeneous class of events, of the kind that you can calculate averages over; like human deaths, they are usually one-off events with their own particular causes.

Other social scientists might have made a little more progress by looking at the social and anthropological aspects of the debt relationship in modern societies. The specific subset of economists who concentrate on finance have analysed the mechanics and consequences of default with the whole toolkit of game theory and of course the lawyers have done all the work they can bill for on the subtleties of bankruptcy law. I spent the best part of a decade learning the ins and outs of this system, but I might not have needed to bother; you can understand a lot of what's important about debt in a couple of paragraphs if you start by considering it as a technology of information and control.

I once got on the wrong side of the late anthropologist David Graeber.* He had written a book about debt from an anthropological point of view, and in doing so he had contrasted it to alternative methods of establishing trade and mutual cooperation. One of these traditions was *dzamalag*, a

---

* In fairness to myself, 'the wrong side of David Graeber' was often a densely populated piece of ground.

ceremonial exchange of goods between trading groups, which took place over an evening of singing, dancing and the exchange of sexual favours. In my review of *Debt: The First 5,000 Years*, I somewhat cattily remarked that although this arrangement presumably worked for the Western Australian peoples who practised it, you couldn't run an industrial economy on the basis of having a wife-swapping party every time you wanted to buy a blanket.

I was trying to make the point that debt, considered as a way of managing information, is extremely efficient. The lender doesn't have to get involved in the management of the business project, or even have to take a shareholder's perspective in assessing how valuable the enterprise is. All they need to answer are two questions: 'Is the promised interest rate attractive compensation for the time and risk?' and 'Do I think I will get paid the money back?' As a capitalist with spare funds to invest, the debt contract lets me make dozens and dozens of investments, well beyond my ability to supervise directly – rather like the thermostats in the squirrel farm in Chapter 4.

This is pretty straightforward. From the point of view of the borrower, though, that variety-attenuating power has some rather more subtle qualities. It introduces a new constraint into the 'survival set' – if you aren't able to make the payments, something bad will happen. And every time you make a payment, it reduces your cash on hand. If you only have a small amount of debt, this just becomes one constraint among others. If a system is loaded up with a lot of debt, the need to make the payments becomes a signal that swamps all other sources of information. It becomes impossible to pay attention to anything that doesn't directly help to generate enough cash to ensure continued viability.

And that's not the end of it; it's possible to push things

further. Add even more debt, and the borrower is in a situation where nothing they can do will generate enough cashflow to service the debt. At this point, debt becomes an instrument of control. The creditor has the option either to break up the company and sell its parts, or to extend another loan to pay the first one back. These powers can be exercised or withheld, in return for whatever concessions he or she can negotiate.

This relationship might be relatively benign – a bank extending an overdraft to a small business that's growing, for instance, so it can reinvest its profits in stock and working capital. Or it might become hostile, with creditors demanding resignations from management and the sale of assets to allow the company to survive. In either case, though, the implicit threat is there – if the creditor isn't paid, they have the right to take actions which have a drastic effect on the viability of the system. That's what makes debt a technology of control.

There are a few more subtleties. Most obviously, it must be taken into account that 'the creditor' isn't necessarily a single entity. If you can find another capitalist to lend you the money to pay back the original one, you might be able to get a better deal, and a lot of the skill of financiers is used in finding clever ways to play capitalists off against each other. Also, I've been intentionally ambiguous about what happens when a borrower can't pay. Differences in bankruptcy codes and legal systems make a huge difference to the relationship and the transfer of information and control. These issues are easier to understand, though, after we've considered how debt worked as a tool of information and control in the second decade of the Friedman doctrine.

How companies lose their minds

When the Covid-19 pandemic hit the nursing home industry, it became, for a short while, impossible to ignore the scandals. Care homes were systematically understaffed; workers had no sick pay and were being made to travel from home to home, acting as vectors of transmission. What was even more curious was that despite the large amount of money spent on provision, and despite all the measures taken to cut costs to the minimum and beyond, care homes did not seem to be profitable; in 2022, they kept going bankrupt, with unbearably traumatic consequences for their residents.

They weren't the only kind of business that seemed to have a mysterious drain on their cash flows. There is practically a subgenre of memoirs today, in which journalists return from New York or London to their childhood homes and write elegiac accounts of the decline of the local employer, and its consequences in terms of poverty, drug addiction and crime in a previously thriving community. A once profitable company keeps laying people off and cutting wages, but never seems to get any more profitable. The executives, who were once the foundations of local civic society, get replaced by interchangeable consultants. Famous brands are allowed to languish for lack of investment, and then sold off overseas.

All these stories tend to lead to one place: private equity and the leveraged buyout phenomenon.

The mechanics of a leveraged buyout are simple enough to have people thinking, 'There's got to be more to it than that.' There's also one key stage to the operation that tends to elicit the response, 'That doesn't sound like it ought to be allowed.'

Basically, you put up a small amount of your own money and take on a lot of debt – the ratio is often as high as 90:10 – to buy the company from its owners. You will be able to borrow most of the debt cheaply, because your lenders know they're going to get paid back quickly. How? Well, once you own the company, you control it, so you can cause it to borrow money. The company you've just acquired borrows money from its bankers (or from the bond market), and you use the loan proceeds to pay back your own creditors.

So there are three steps. You begin with zero assets and zero liabilities. There's an intermediate step where your assets are a company and your liabilities are some debt. And there's a final stage, where you're free of liabilities, but your assets include an indebted company. This is the bit that gets ordinary non-financial people worrying – you're effectively using the company's own cash and assets to pay for your acquisition of it.

There are a number of advantages to doing things this way. Most obviously, you greatly increase your potential returns on the money you invest. If things go well, then the debt which the company took on gets paid back over time, and ten years later you might find yourself owning all the assets, free and clear of debt, having only put up 10 per cent of the cost at the first stage. There are also tax benefits, because debt interest is deductible from taxable profits while dividends to shareholders aren't. By taking on the debt, you can turn ten years' worth of profits which might otherwise have been taxable income into a long-term capital gain, which is often taxed at a more favourable rate.

Set against this, the big disadvantage is that taking on a load of debt might cause the company to go bankrupt. But this isn't such a risk to you as an investor. The company has limited liability, so the most you can lose as the buyer is the initial 10 per cent that you put in. In fact, it's perfectly possible to come away with a personal profit from a deal in which the company went bust, if you only put up 10 per cent of the money and prioritised paying yourself dividends rather than reducing the corporate debt. The company's creditors can't chase you for the rest, because of limited liability. But they can take away valuable assets and buildings; they will usually be all right.

The people who are really at risk, though, are the ones who depend on the company continuing to exist as an ongoing entity – workers, suppliers, customers and managers. The little people.

And here's the trick of the thing – this big disadvantage is actually the big advantage. The leveraged buyout completely rewrites the incentives of the management system. All other priorities have to be subordinated to that of ensuring that there is enough cash to continue to survive. Rather than appealing to the sense of duty that the executives might or might not feel to shareholders, the debt burden creates an ultimatum – if they don't concentrate their efforts on increasing profits, the company will go bust. And the management of a bust company might never get an equivalently prestigious and well-paid job again.

So, if you have a company where the managers are extracting perks for themselves – private corporate jets, country club memberships and ridiculous expense accounts, for instance – a leveraged buyout will stop them doing that. It works even better if the managers are being lazy, not taking difficult decisions, allowing obsolete product lines to plod on or becoming dependent on back-scratching relationships with single large customers who are equally sclerotic. It works best of all if you have a company that has managed to confuse itself about what it ought to be doing, thanks to the hallucinations created by a really bad accounting system.

Finding those kinds of companies was meant to be the skill of the buyout firms. From the 1980s onward, while the Friedman doctrine was the ideology of corporate America, the leveraged buyout boom was the way in which it was enforced. Specialist bankers and financiers, working with pools of private capital, scoured the country looking for sclerotic, wasteful or inefficient companies, then buying them up and making a fortune, or letting them fail and moving on to the next. The 'private equity' industry, as the leveraged buyout financiers liked to be called, was an extremely efficient predator – compared to all the other non-human decision-making systems, it was much less slow-moving and very aggressive.

Part of the reason for this was that it had what many other institutions lacked – a very strong sense of identity and ideology, a 'System 5' capacity that allowed it to integrate its information gathering and operational management functions with a single clearly defined goal. That ideology had started with Friedman, but it was developed by others. If you want examples, there's one publication that stands out among the rest.

The most ideological academic journal of them all

There is one exception, of which I'm aware at least, to the general rule that the purpose of academic journals is to generate citations. The *Journal of Applied Corporate Finance* was set up in 1988. Having read almost every issue of the first ten years of its publication, I look back on it as one of the most explicitly and thoroughly ideological publications I've ever read – including those printed by the Soviet or Chinese Communist Party.

Along with the theoretical and mathematical articles, the *JACF* used to print case studies. Unlike most academic economics journals, it was clearly written and interesting; quite often it would carry edited and abridged versions of articles from more technical journals. Consequently, it read like a book of kids' stories, each one having a moral purpose in which virtue triumphed and a lesson was learned. It achieved this effect, I think, because it had a consistent and clear philosophy.

Some of the article titles are perfect time capsules. Picking up the 1992 volumes, you can read 'Financial Innovation and Economic Performance' by Robert C. Merton (his hedge fund, Long Term Capital Management, went bust six years later). There's also 'What Pay for Performance Looks Like: The Case of Michael Eisner' (his turbulent period as CEO of the Walt Disney Company was later written up in a book called *Disney-War*). But the one that sticks in my mind is a study called 'The Market Rewards Promising R&D – and Punishes the Rest'. You can really imagine that as a slogan painted on a wall in a revolutionary society, and in a sense that's the role it was meant to have.

The overriding intellect was that of Michael Jensen, a financial economist who was largely responsible for bringing the Friedman doctrine on to a rigorous mathematical basis, the better to integrate it into the governing worldview of economics. He wrote an article (with William Meckling) called 'Theory of the Firm: Managerial Behaviour, Agency Costs and Ownership Structure', which appears to have been cited in nearly every issue of the *JACF*.

You can trace the scientific translation of the Friedman doctrine just by looking at each element of the subtitle of Jensen's article. 'Managerial behaviour' refers to slacking, not making tough decisions and potentially pursuing social goals with the shareholders' money. 'Agency costs' refers to the difficulties experienced by shareholders in checking up on managers. And 'ownership structure' refers to the solution of the previous two problems – the use of debt and stock options to ensure that managers act as if the corporation didn't exist, and as if they had been hired directly by a personal owner who was constantly looking over their shoulder.

The intellectual backing for the leveraged buyout movement was explicitly disciplinary, and the use of debt as both a technology of control and as a way of serving the interests of capital-owners is clear as the moral of the case studies. Companies would pay out huge dividends and take on self-consciously risky amounts of debt in order to 'create a sense of urgency' among their management, or to communicate management's confidence that their accounting policies weren't as aggressive as they looked.* Leveraged buyouts were given 'reviews' in the journal as if they were musicals or restaurants, and judged according to how aggressively the management changes were made.

The *JACF* was also concerned with promoting the idea of financial markets as computing fabric. Jensen himself said, 'I believe there is no other proposition in economics which has more solid empirical evidence supporting it than the efficient market hypothesis' (the proposition that the share price of a company summarises all available information about the

---

* Specifically, Sealed Air Corporation (the makers of those plastic inflated packaging things used to ship delicate objects) paid out dividends to create a sense of urgency, and in fairness, they got away with it; the CEO credits the move with making it possible to extend workers' share ownership and with improving the company's performance to this day. The company taking on debt to communicate confidence in its accounting policies was called CUC International, and it's more embarrassing for the journal – five years after the article was published, it became the largest accounting fraud that the US government's Securities and Exchange Commission had prosecuted up to 1998.

financial future of that company at any given time).* And this idea provided the moral homily to many of the case studies; the authors would demonstrate that having followed the advice of modern applied corporate finance, the managers saw their share price rise, demonstrating that they had done well and been rewarded.

It was only one journal among many, but it was on the syllabus when I was at business school for a reason; that was the way to think if you wanted to get on in finance or consultancy. A substantial proportion of the case studies and articles applied the 'Economic Value Added' model invented by Joel Stern, the founder of the management consultancy Stern Stewart.† This was basically an application of the finance theory of the 1970s to reinforce the management theory of the 1980s; it set a financial benchmark for investment returns, to determine what should be done and what should be cut.

As is typical in the consultancy industry, the Economic Value Added model was so successful that all the competitors invented their own version, allowing consultants to carry out the equivalent of leveraged buyout management in companies which

---

* In other words, literally, that the stock market is a computing fabric. The argument can be sketched quite easily – if there was some additional information, it would either be positive or negative for the share price. So, somebody would have the incentive to find it out, and make a profit by buying or selling the shares. And in doing so, they would move the share price to where it would be if the piece of information was generally known. It's an application to the stock market of the principle summarised in the joke about an economist assuming that a $20 bill lying on the sidewalk must be an illusion because if it was really there somebody would have picked it up.
† Stern Stewart sponsored the journal for a while, as did the investment bank Morgan Stanley; it's currently owned by a private equity guy.

weren't suited to actual corporate raiding. Business schools and consultancies, as we mentioned earlier, are a key part of the internal communication system of management, and journals like this are how that communication system aligns itself. The *JACF* and its ideological peers were read by the people who went into the consultancies, banks and investment firms – they all shared the same language and the same assumptions.

That's part of what made them so fearsome and effective, like any well-drilled cadre. Across the economy, corporate managers were left pleading about their history and community, the trust of their customers, the value of their technology. And time after time, they found out that unless it could be quantified in an accounting system and placed into exactly the right kind of spreadsheet, this kind of information no longer mattered. So they borrowed more money, paid bigger dividends and hoped.

How it began to go wrong

The trouble is that things drift. The leveraged buyout cadre started in a target-rich environment. In the 1970s, it had been tacitly accepted in plenty of companies that the era of capitalism was over and the managerialist system was the way of the world. Executives were able to draw rewards similar to those of ownership by paying themselves generously, using expense accounts to provide them with luxury goods and gradually filling the ranks of middle management with friends and family members. The cash came in, some of it could be spent on speculative future projects and the rest guzzled, with an occasional dividend paid to the shareholders for a sense of propriety.

These companies made great investments for leveraged buyout firms. But once you've done all the good deals, what do you do then?

A true believer in the ideology might have wound up the fund, distributed the cash to the investors and declared that there were no more sufficiently attractive investments to make in this particular style. The investors could have taken the money and gone off to look for new opportunities in emerging markets or Silicon Valley start-ups, remembering the leveraged buyout boom as a great money-making episode and a salutary punishment operation to remind the managerial class who their betters were.

But who would have had the willpower to do that? The alternative course of action was to simply expand the definition of companies that could be considered valid targets. Rather than looking for inefficient practices or self-dealing executives, you might just start searching for companies that weren't making full use of the tax benefits of debt finance. Or anything where the cashflow was stable enough to get a cheap deal on debt. Or you could just take advantage of a temporary dip in the stock market to get an attractive entry price. And so the practice continued; the buyout sector continued to grow, and companies started to pre-emptively act as if they were vulnerable to take-over, in order to preserve themselves from becoming targets.

From a cybernetic point of view, it's interesting as an example of how the systems and structures mattered so much more than the individuals involved. The development of the Friedman doctrine into the intellectual backing for the leveraged buyout boom and the private equity industry are best seen as a conflict between two comprehensive systems of interest, both of which might have regarded the other as a threat. The great unremarked class struggle that happened in the 1970s and 1980s was that between capitalism and managerialism.

The managers lost this struggle, pretty comprehensively. And as we've seen, the combination of the blind spots in management and the blind spots in economics came together to produce an ideology which was bound to remove management capacity. And that created further blind spots, and further reduced the system's ability to cope with shocks. The story of how we got to where we are is a story of the attempts of the system to cope with this, and to search for short-term equilibrium.

9

The Morbid Symptoms

The crisis consists precisely in the fact that the old is
dying and the new cannot be born; in this interregnum a
great variety of morbid symptoms appear.

Antonio Gramsci, *Prison Notebooks*,
trans. Quintin Hoare, 1971

I wanted to write a sort of cybernetic political thriller, but it
didn't quite work out that way. It seems that in order to get to
the point where the story begins, you need to write eight chapters explaining the construction of the murder weapon. But
here we are. To recapitulate, the basic problem is that systems
in general need to have mechanisms to reorganise themselves
when the complexity of their environment gets too much to
bear. But the high-level governing systems of the industrial
world – economic policy and business management – had some
defects and blind spots which prevented this from happening.
How did that turn into an economic crisis, and how did it
mature into an ongoing political polycrisis? How did we get
from the Friedman doctrine and leveraged buyouts to Trump,
Brexit and populism?

The first part of the story involves expanding on the conclusion of the last chapter. A key part of recent history has been
a story of class struggle, but not the usual kind. There's been a

class war between the capital-owning class and the managerial class. It might be easiest to illustrate this by looking at someone who started off on one side of this conflict, but quickly turned his coat.

Class traitor to the managers

Jack Welch, the CEO of General Electric from 1981 to 2001, once called shareholder value 'the dumbest idea in the world'. But he said that in 2009, six months after the collapse of Lehman Brothers and at a time when lots of people were revisiting points of view they had held in the past but were now less keen to be associated with. In his career, while he could really make a difference, he was the king of shareholder value.

Welch's rise to prominence as America's most admired CEO (indeed, in 1999 *Fortune* magazine crowned him 'manager of the century') and the ubiquity of his grinning face in magazines was based on the extent to which GE's share price had kept rising while he was in charge. His management style reflected the consequences and pathologies of the fully realised Friedman doctrine and Economic Value Added. Welch's GE was a company which always 'made its numbers' – the quarterly earnings per share would always meet or exceed the forecasts of investors.

After he left, people questioned whether his success was legitimate. General Electric's financial subsidiaries grew so big under his tenure that they were more important to the company's earnings than any of the manufacturing operations, and the earnings of financial services businesses are dependent on subjective estimates about the future. But even without that shift, it's likely that Welch would have ensured that GE made its numbers, simply by making it a priority to do so. He institutionalised this practice,* ignoring the fact that a quarterly financial accounting number – or even a few years' worth of them strung together – is an extraordinarily crude information filter.

One consequence of focusing on quarterly numbers and short-term delivery is that costs become more important than revenues. To increase its revenue, an industrial company like GE would have to find a customer, win the order, make the product and get paid – hard to do in twelve months, let alone three. An employee can be fired more quickly than that; a business can be shut down or sold. The management has greater control over costs than revenue – they depend less on decisions taken outside the organisation's perimeter.

Welch ended up with the nickname 'Neutron Jack', after a bomb that allegedly left buildings standing but vaporised people. He closed down parts of the business according to a principle that GE would not remain in a market unless it had a good prospect of being either global number one or two. And if that wasn't enough, he instituted a 'rank and yank' policy. Every year, managers were assessed, and the bottom 10 per cent were fired regardless of their absolute performance. Further cuts were made wherever there was perceived to be dead wood, with whole layers of middle management removed in a constant effort to cut operating costs.

Was this the right thing to do? Maybe, at the start of Jack Welch's time, GE had loads of dead wood. Maybe it needed to cut costs, get out of businesses that were going nowhere

---

* As with the comment about shareholder value, Welch didn't always care to be identified with this practice, but his chief financial officer gave it away in interviews, saying things like, 'Of course we're buying earnings when we do an acquisition,' and, 'I see nothing wrong with someone saying, "Look, I have an earnings objective for the year."'

and institute a management structure that felt like a portfolio of start-ups, each focused on delivering their own quarterly numbers. Since the financial sector was massively expanding, maybe it wasn't such a dumb decision for GE to do the same thing, even if it did blow up slightly in the aftermath of Welch's retirement and throw a pall over his legacy.

The point here is not to get into an argument with the ghost of Jack Welch about shareholder value – like Milton Friedman, he's representative of a culture. The point is to consider the question: as management practices like this became embedded in industry, what effect might this have had on managers?

Revolutionary capitalism

Jack Welch became chief executive of GE ten years after the publication of the Friedman Doctrine essay, just as the leveraged buyout boom was really getting started. His tenure was a particular cultural moment in American business; the financial magazines called it the 'shareholder value revolution'. What happened to the managerial class was more or less the same thing that happens in every revolution or coup – a few people prospered mightily, some 'disappeared' and the majority got used to living with the new system. Jack Welch was in the first category; he ended his career as a well-established member of the capital-owning class and one of the richest people in the world. Stock options, equity grants and greatly increased salaries and bonuses were important to the shareholder value movement, and plenty of articles in the *JACF* made the case that what looked like truly rapacious pay packages were in fact a good deal for the shareholders.

The question of whether the shareholders got good value for money is not so important for our purposes. Plenty of people will argue that they didn't, that there's little evidence that higher pay resulted in better performance. Cynics claim that the executive compensation system was a racket, put together by a clique of cronies who sat on each other's committees and approved each other's bonuses. But whatever the truth of that accusation, it was one of the ways that the system reorganised – how the message was communicated that the managers were in alliance with the capitalists. Just as was intended, the expansion in stock options aligned the incentives of managers with the owners.

Lower down the scale, lots of managers stopped being managers. Stock options were the reward for generating short-term profits, but employment insecurity was the other half of the deal. The threat of redundancy performed the same function at middle management levels as the judicious use of debt at higher levels – it swamped the decision-making processes with a requirement that targets had to be met. For Welch, this disciplinary effect of removing layers of management seems to have been at least as important as the cash savings; that's why the rank-and-yank of the bottom 10 per cent was carried out to preserve the threat, even when economic conditions made it hard to justify.

Removing middle managers obviously reduced the cognitive capacity of the organisation, but this shouldn't necessarily have resulted in a degradation of the capability of the system as a whole. After all, those managers didn't cease to exist; the net increase in the size of the professional and managerial class over the period in question was substantial. Where problems were created, it was by the reorganisation of the system – by the managers who stayed and adapted.

We've already seen how accounting systems can act as an information-reducing filter. It follows from this that the greater the emphasis placed on accounting-based targets set by the CEO, the greater the filtering effect. The progress of financialisation exacerbated this in two ways. First, it cemented the dominance of financial reporting over any other kind of information system. Rather than making a resource bargain between the operating and strategic management and tailoring it to the specific requirements of different processes, every unit had to contribute to a smooth progression of quarterly numbers, often with absurd effects.

In 1997, for example, GE closed a transaction with a defence contractor which gave them a one-off gain of more than $2bn. But recognising such a big spike in profits would have made it difficult to continue to show steady growth in future years. So they closed down some factories, making more than 1,000 people redundant, in order to generate a corresponding one-off cost. 'Offsetting the gain has been our practice,' was the comment at the time from Welch's spokesperson.

During the housing bubble of the 2000s, GE had become extremely reliant on its financial services arm, which was one of the biggest subprime mortgage lenders in the USA. In order to make the numbers (in the words of an internal auditor quoted in a later press release announcing a $1.5bn settlement with the Justice Department), they 'jacked up the volume without controls', and misrepresented the quality of their lending to investors they sold the mortgages to. If you consistently demand the impossible, you will inevitably get the unethical.

And second, as we've seen in the previous chapter, the use of debt in the overall system meant that across the economy, the resource bargain between top-level management and the operational businesses was further constrained by the need to generate cash. This is how we get stories of neglect at nursing homes, underinvestment and decay in manufacturing, and the gradual decline of prosperous industrial towns into drugs and poverty. But the real disaster for the information processing capacity of the capitalist economy was outsourcing.

Serious problems and no-brainers

One of Stafford Beer's key criteria for establishing a viable system was that careful attention needs to be paid to preserving information when a signal crosses a boundary – the 'translation and transduction' problem. There is always the issue of ensuring·that information is received in a form and at a time which allows it to be part of the decision-making process, but it is also the case that communication is just difficult. As anyone who has tried to organise a moderately complicated set of meetings between people at different organisations knows, cross-purposes and misunderstandings are frighteningly common.

A lot of the seeming redundancy in middle management used to be dedicated to mitigating this problem; extra capacity and processing power was installed at the communication boundaries, to make sure messages got through with the appropriate degree of nuance and content, and that wobbles and flutters would be handled at the appropriate level. This is one of the things that gets thrown away when companies outsource. By looking for the activities in which they could be global leaders and outsourcing everything else, companies exchanged internal boundaries for external ones. Since a large part of the reason for doing so was to economise on management capacity, these relationships necessarily attenuated information – that was the purpose of doing so.

Outsourcing is a contractual relationship, and contracts (like debts) are typically all-or-nothing affairs. Either a requirement has been fulfilled or it hasn't. The typical outsourcing contract narrows the bandwidth of the communication channel to either 'everything is going more or less as anticipated' or 'it's stopped working and we need to find out why'. If it's the first of these two cases, everything's fine. If it's the second, the continued stability of the system will depend on how much information-handling capacity can be brought to bear to address whatever problem has arisen.

Since the outsourcing communication channel is designed to spend most of its time transmitting 'everything's OK', it's difficult to guess how much additional bandwidth needs to be allocated to carry messages like 'but the following conditions are changing which might affect things in the near future' – let alone how much might need to be allocated at short notice when it starts to say 'everything's no longer OK'. If you have targets to make, it will always be tempting to cut out a bit of spare capacity.

And this is of course what happened; in an accounting system which targeted overhead costs, combined with an emphasis on generating cash, it looked like a no-brainer to thin out the ranks of middle managers who didn't appear to do anything. But unfortunately, a 'no-brainer' was exactly what it turned out to be. Over the course of Jack Welch's career, the industrial world's productive system – the corporations – set about the equivalent of amateur brain surgery, hacking away bits of their regulatory and information-handling system, to see if they could do without them.

How did they get away with it for so long? By breaking another implicit deal. What is the capitalist system *for*?

The broken covenant

In 1914, a man called Jerome Levy sat down to think about the economy. A physicist by education and a wholesaler of knitwear by profession, he wanted to answer a question that often strikes the minds of businesspeople. He ran his firm with the objective of roughly a 10 per cent return; he would spend $1,000 on inventory, rent, salaries, electricity, taxes and so on, and expect to get $1,100 back. So where did the $100 come from?

Like many good questions, this one is deceptively simple. At one level, it's the mark-up on the cost of goods sold, reflecting the value added by the service of wholesaling, stockholding and distribution. This answer is good enough for most chamber of commerce members, but Levy's mind was a little more curious. In aggregate, the companies of America made a profit every year; the total amount spent in acquiring things and paying salaries came back, but so did a bit more. If you'd been educated in laws of the conservation of mass and energy, this might seem like a puzzle.

It is actually a puzzle, and one that has fooled smart people, including a significant minority of living PhD economists. A broad-brush answer might be to note that the number of dollars in the economy isn't fixed and shouldn't be expected to obey any conservation law – they can be created and destroyed at zero cost. The amount of dollars which needs to be created depends on the flow of transactions, and if that's growing, the monetary system will reflect that. The relationship between dollars and transactions is itself not straightforward; transactions happen all the time, and are bundled up into arbitrary reporting periods by the accounting system.

These somewhat general explanations give the flavour of the thing. What Jerome Levy did was to write out a set of equations which defined such an accounting system; every dollar came from somewhere and went somewhere, the creation of dollars by book entries was consistent with the transactions, and the quantity of profit was consistently explained.

The system of profit equations that Jerome Levy wrote down in 1914 anticipated a similar set of equations written down by the Polish economist Michal Kalecki in 1935. And Kalecki's system is regarded by a lot of people as containing nearly all of what's useful in J. M. Keynes' *General Theory of Employment, Interest and Money*, published in 1936 and widely accepted as one of the greatest works of economics ever.

Levy went on to demonstrate that the proverb 'if you're so smart, why aren't you rich?' was not applicable in this case; aided by his sons, the Levy family went into finance with sufficient success that the Jerome Levy Forecasting Institute they endowed at Bard College continues to promote their approach to economics today. You used to be able to buy a copy of the book Jerome wrote in 1943, *Economics Is an Exact Science*, from them; I got mine in about 2002.

In the book, Levy sets out his view of the purpose of capitalism:

> The working class is the original and fundamental economic class . . . The function of the investing class is to serve the members of the working class by insuring them against loss and by providing them with desired goods. The justification for the existence of the investing class is the service it renders the working class, measured in terms of wages and desired goods. The contrary is not true. The working class does not exist to serve the investing class. The working class has the right to insure itself through organizations composed of its members or through government, thereby eliminating the investing class.

It almost seems shocking to see that written down in so many words, but the idea that the investing class might have a purpose or obligations was reasonably common before the ascendancy of the Friedman doctrine. Hank Greenberg, the founder of American International Group, used to say that he would never use redundancies as a means of managing the business cycle; this was the implicit promise that AIG made in return for demanding that its executives sacrifice so much of their lives to the company.

That idea was reflective of a long-term bargain between firm and employee, with both sides expected to make investments and plans but one in which the capital side bore most of the risk. In other words, exactly the kind of relationship that justifies the existence of a long-lived limited liability company rather than a spot market, and one which doesn't fit terribly well into the standard economic model.* One partial working definition of 'neoliberalism' might be that it is the ideology which denies the existence of this relationship.

Reversing the polarity

In fact, the neoliberal era saw this bargain reversed; if the investing class was capable of providing business cycle insurance to the working class, then you can guess it might be possible to run the system in the opposite direction. Slightly ironically, the

---

* This point needs to be reiterated every time it comes up: economics *can* handle all sorts of concepts and relationships, but 'tractability' becomes a huge issue in modelling; when economists want to do applied work, they have the choice of a tricky model that's designed specifically for a single case or making a lot of simplifying assumptions. Even if they choose difficulty and realism, this might not help as much as you'd hope, since when theoretical models are practically applied, they're often reduced to a system of linear equations.

first person to realise this was Karl Marx, who wrote about the 'reserve army of the unemployed', and their role in managing wage costs by reducing the bargaining power of workers. Michal Kalecki's version of the Levy profit equations could be interpreted as showing a relationship between the reserve army and wage inflation, and so could Keynes' general theory.

But it was a conclusion that could be reached from several directions. Milton Friedman had a model of the 'natural rate of unemployment', which would cause inflation if it was exceeded, and this was refined over the years into the NAIRU (non-accelerating inflation rate of unemployment) which formed the basis of monetary policy under the nineteen years' stewardship of Alan Greenspan at the Federal Reserve.

Although unemployment is, in one sense, a waste, the alternative possibility is for managers and capital-owners to lose control of the wage bargain. For that reason, the capitalist class is often prepared to tolerate lower output as the price of retaining control of wages and prices. The secret to the Great Moderation of the neoliberal era – the sharp decline in the volatility of output and inflation – was that economic policy began to use the unemployment rate as a variety attenuator; external shocks could be accommodated or passed on, so long as wage pressure and price inflation did not build up.

This was why the corporate management system of Jack Welch required the economic management system of the Great Moderation. The global economic system had gradually removed vital parts of its own brain; informal and rich communication channels within organisations were replaced with formal and contractual links which often crossed time zones. Intelligence functions that were meant to consider the future were swamped with debt, so that they couldn't pay attention to anything that didn't generate cash for the next payment.

Because it was reducing its capacity, the corporate system with its managers and investors was no longer able to absorb volatility and shocks; the economy had to be reorganised so the working class performed this function. It looked like the final triumph of capital, and it was.

Until it wasn't. We discussed in Chapter 5 how the blind spots and weaknesses in the overall system of governance were so placed as to miss the debt bubble and consequent global polycrisis. This bubble was, like almost everything that goes wrong in a cybernetic system, a solution that turned into a problem.

The road to perdition

A funny thing about ideology is that it's difficult to confine it to where it's useful. There was not much reason why the political management system of the state should have reorganised itself along similar lines to the economic management system of the corporations, but it did. The same professional and managerial class runs both, so ideas spread naturally from one to the other. And, as can be seen from Milton Friedman's original 1970 essay, his doctrine in corporate management was part of a general political project. It was easy to apply the same concepts of efficiency and the removal of dead-wood bureaucracy to state organisations. Governments were also subject to the same cognitive pressures of a world that was becoming more complicated, so they also needed solutions to attenuate information.

But the public sector lacks any direct equivalents of the debt relationship and the leveraged buyout. Countries can borrow money, and they can default on their debts, but bankruptcy doesn't carry the same consequences as it does for private sector entities; there's no transfer of control rights. When dealing with the governments of poor countries, there is some scope to use the debt relationship as a tool of control, simply by making a credible threat not to lend any more money – that's how the IMF operates. But for a country that's capable of borrowing in its own currency, this disciplinary aspect just isn't there.

Nor is there any single maximising goal that corresponds to shareholder value; the nearest equivalent might be the minimisation of total public expenditure, which seems to have been the performance measure that many government reformers latched on to. Although you can play at markets and pretend to create competition, the systems are fundamentally different. The imposition of Friedmanism and the debt burden in the corporate sector gave a clear signal about the new priorities, but the equivalent processes in the public sector just added a lot of noise. The system was calling for 'reform' and revolution, but not giving any clues as what specifically ought to be done.

And so the progress of outsourcing and decerebration ended up being even more intense, and the public sector became as big a fee pool as the private sector for the consultancy industry. This had the effect of closing off the other possible way in which the working class might have insured themselves against volatility – through the policies of governments that they elected.

The public sector atrophies

One consequence of the apparent success of the delegation of interest rate policy to central banks was that people started to look for other areas where something similar could be done, creating an accountability sink to improve policy outcomes. It was also attractive for governments to do things which helped them negotiate one of the fundamental constraints of democratic politics: the inability to make decisions which bound the next government. Creating an arms-length agency to administer a policy area provided an opportunity to institutionalise a particular set of priorities; even if you lost the next election, dismantling that institution would be difficult. Similarly, the accounting system could be exploited to lock in expenditure by making long-dated contracts with outside providers. Although the ratio of government expenditure to national income continued to grow over the Great Moderation, the proportion of it which could be considered attributable to identifiable politicians shrunk.

The decerebration of the public sector happened in a similar way to that of the private sector; they used management consultants for short-term injections of brainpower, and outsourcing contracts which had the effect of removing management capacity in the long term. But the consequences were, if anything, more severe. Corporations, as a rule, had cut away peripheral activities in order to save capacity for their core competence. The state, when it outsourced, often did so with respect to absolutely core functions, like running trains, building hospitals and even managing its defence procurement. The precedent was set early on by central bank independence; few things are more central to the powers of a state than the value of the currency, so if that can be delegated to technocratic outside bodies, what can't?

This had a frightful effect on public sector management. The coordination function was impaired; the difficulty of 'joined-up government' and making policy for problems that crossed the boundaries of different agencies was repeatedly remarked on. And the operational delivery functions started to suffer severe cognitive loss, too. A company that sells goods and services for profit can never completely sever the connection which takes information from its customers; the people who buy the thing have the ability to refuse to do so. In many cases, people who interact with the state don't have even have the ability to transmit that single bit of information because they can't shop elsewhere; they can complain if they like, but they interact with the service representative, the paradigmatic accountability sink.

Things got worse over a long period of time, but this was initially hard to notice. Recall that in Jerome Levy's high-level view of the economy, the investing class has two purposes – providing insurance against the business cycle to the working class, and providing them with consumer goods. While the first of these services had been abandoned, this was not immediately obvious – the business cycle itself had been temporarily calmed down. And although many of the purchases were funded by debt, the second still seemed to be functioning. Over the course of a few decades, the risk transfer was completed.

The wizards behind the curtain

The financial crisis of the late 2000s is the sort of event that people who feel the need for clean breaks are bound to be attracted to; if you want to identify a point where the neoliberal era came to an end, that seems like a good place to draw the line. Not everyone put it quite as pungently as the Irish economist Colm McCarthy, but his reminiscence a few years later captured a general shift in sentiment. As he said, 'Everyone in Ireland had the idea that somewhere . . . there was a little wise old man who was in charge of the money . . . They saw him and said, "Who the fuck was that? Is that the fucking guy who is in charge of the money?" That's when everyone panicked.'

That may describe the panic, but it's also the story of the

preceding two decades. People were horrified to see that the chief executive of the Irish Financial Regulator was a normal little man with a moustache and a desperately unconvincing manner in television interviews, just as Dorothy was surprised to draw back the curtain and discover the Great Oz. But it wouldn't have been shocking to discover that central bankers were just normal civil servants if they hadn't been treated as wizards for so long.

It all started, as one might expect, with an accountability sink. Central bank independence wasn't really a new thing; the United States Federal Reserve had possessed autonomy since its formation, and the Bank of England had been a private company with shareholders until it was nationalised in 1945. But monetary authorities had mostly recognised that their independence was something that was permitted by the democratic government, rather than a power they exercised by right.* This view began to be challenged in the 1970s, as the economists Edward C. Prescott and Finn E. Kydland proved (in their model) that democratic control of the money supply was an unaffordable luxury.

The broad structure of the argument was what they called 'time inconsistency'. The idea was that governments knew what they should do in the long term, but that they were subject to destabilising feedback from the election cycle. Without a great deal of evidence, economists asserted that in countries where monetary policy was controlled by the political authorities, they would always be tempted to create an unsustainable

---

* The Deutsche Bundesbank is a possible exception to this; its independence was guaranteed by the post-war constitution, and it regularly made it clear that it regarded itself as having greater legitimacy than any potentially fleeting government.

mini-boom in an election year, trading short-term inflation for short-term growth in wages and employment.

But since the artificial people in the economists' models were assumed to be rational and far-sighted, there was not really any such thing as short-term inflation or a trade-off between employment and prices; people would build the political cycle into their wage demands and pricing. The outcome was therefore that the overall level of inflation would be higher at all times and anti-inflationary policy would be less effective.

In this model, the government is trapped by its short-term needs in a self-destructive cycle. Everyone can see that the policy is disastrous over the long term; inflation would build up to such a degree that much more drastic action would have to be taken. But at any moment before the disaster arrives, it is less painful to go through one more cycle of obtaining easy money. The solution, in the model, was for the government to hand over control to an independent, unaccountable body, the equivalent of Odysseus tying himself to the mast to avoid being tempted by the sirens. (This metaphor was done to death during the period in question.)

In cybernetic terms, the central bankers of the period seem to have concluded that the problem with the system was that the channels of accountability from those affected by unemployment and recession were working too well; they were transmitting too much information and impairing the ability of the system to manage itself. This might seem like a curious conclusion now, but hindsight obscures the concern over inflation and volatility. There was a powerful sense in the air that something had to be done so that the inflation of the 1970s was not repeated.

Finance ministries took the critique on board; throughout the 1980s and 1990s, monetary authorities became less and less responsive to outside influence. When the Berlin Wall fell, the new capitalist economies were advised to each set up an institutionally independent central bank to manage their monetary policy. And then, in 1998, eleven of Europe's industrialised economies handed over control to the European Central Bank with the founding of the Euro. The founding treaty of the ECB ensured that it would be an institution with more powerful safeguards over its independence than any other seen before.

And it worked, for a while. We've seen that using the working class to absorb variety was successful in the short term, but it needs to be emphasised: the new policy architecture really looked like it had worked. If you sit down in a library and go through the publications of any central bank for the years 1998 to 2007, you'll see many articles claiming that things were going better than they had ever gone before. And you hardly even need to make this effort; look at a chart of inflation or GDP growth for anywhere with an independent central bank, and you can do intraocular trauma statistics (it hits you between the eyes). There was year after year of steady growth with low inflation.

When the crises began, it was because of blind spots in the system. The level of overall debt ought to have been managed to keep it in a survivable region. But instead, only government deficits were monitored; private-sector debt was outside what the control system was monitoring. It wasn't closely monitored by the economic policy framework, and it was regarded as a positive thing by the post-Friedman system of corporate control. And it turned out that this fundamental control parameter of the economy was being used to absorb shocks and variability from elsewhere.

Nowhere was this easier to observe than in the American mortgage market. In the face of stagnant real wages, the middle class borrowed money to maintain their consumption levels. This was made possible by falling interest rates and rising house prices, which could be seen as evidence of successful management by the central bank – if you ignored the debt numbers.

Then the crises began.

El pueblo

When Stafford Beer was making his initial presentation to President Allende, he drew his diagrams and built up the components of the viable system model, showing how basic operational systems were embedded in larger blocks for the purpose of coordinating their activities. These management units were in turn embedded in the economic sectors for the purpose of agreeing the resource bargains which constituted the national economic plan. The plan was itself a bargain, struck between the systems responsible for optimisation of current production and those that looked out to the future. And, as we've seen, balancing those two systems and managing the development of the bargain between them was the responsibility of the highest-level function in the model, System 5. As Beer tells the story:

> I drew the square on the piece of paper . . . [The President] threw himself back in his chair: 'At last', he said, 'el pueblo.' This remark, as I have previously attested, had a profound effect on me.

The people. It's hard to know to what extent this was a rhetorical flourish on the part of Salvador Allende, but one of the most underestimated techniques of political and social analysis is to look at people's jokes and metaphors, and take them literally.

One of the big problems with cybernetics is that it's difficult to know when to stop. The web of causal connections and information flows never reaches a tidy end with everything tied up; every system turns out to be a subsystem embedded in a larger system. When you go down the chain, you can eventually come to a stop as the units of analysis reach the size of a single human being. But when you go up, every management unit is itself part of a bigger unit and so on until . . .

El pueblo. The system closes itself – or at least, it does in any non-totalitarian society – by the fact that the highest-level decision-making system operates by consent of the decided-upon. Even in a dictatorship, there is a collective veto capable of being exercised if the situation becomes intolerable to the individuals who make it up. Even the abstract, high-level, unthinkably complex cybernetic entities we've been thinking about – slow-moving artificial intelligences like 'global capitalism' – are embedded in an even slower, even more abstract collective decision-making system of the whole population. We just don't notice its existence, for two reasons.

First, it hardly ever does anything. The purpose of a top-level system – System 5 – is to be the last-resort absorber of information. It is there to deal with the shocks that arise from imbalances, and be ready to adjust itself. If the system is working correctly, all parameters are balanced and there is no excess variety to absorb. When things are going well, you don't notice *el pueblo* as a collective decision-making entity. Things only start to happen when the chips go down.

And second, *el pueblo* usually doesn't have a postal address, let alone a telex link to the presidential palace. In general, hardly any effort is expended on considering what kinds of communication channels should be maintained to allow the population to express its views to the government, or at least, not from an information engineering point of view. It seems quite clear that different arrangements might have different characteristics – a proportional election system should be capable of carrying slightly more information than a first-past-the-post one, a monthly opinion poll has a shorter lag than an annual one, and so on, but this isn't how they're thought of; elections are simply horse races with executive power as the prize, and opinion polls are rarely used as more than a sort of racing form to predict the winners of the next race. The channels seem to be designed to carry very few bits of information.

The only kind of communication that such a constrained channel can carry is a scream: the signal that passes through the levels of control and announces that something has gone wrong which threatens the integrity of the system itself. This is why there was a family resemblance between the 'populist' movements that sprang up in the 2010s. Narendra Modi in India, Beppe Grillo in Italy, Donald Trump in America, Nigel Farage in Britain or Recep Tayyip Erdoğan in Turkey instinctively realised that they were on the same side; each of them, in their own culturally specific context, was acting as a communication channel for a population which wanted to convey a single bit of information: the message that translates as, 'HELP! THE CURRENT STATE OF AFFAIRS IS INTOLERABLE TO ME.'

People have written tens of thousands of words trying to understand what makes Trump voters, Brexit supporters and their local equivalents tick. By and large, they haven't come up with much; all academic anthropologists or newspaper journalists seem to produce is a head-shaking admonishment that the middle-class elites ought to listen to these people, with hardly ever a clue as to what they might find out if they did. The reason why all these quests for the true views of Real America or its geographical equivalents seem to reach such invariably disappointing conclusions is that they're looking in the wrong place. The medium itself is the message; what liberal society ought to be responding to is the *fact* of mass distress, not its content.

Overload and despair

One notable thing that happened during the early months of the Covid-19 pandemic is that the suicide rate in England and the USA went down, despite the fact that many people had confidently predicted that the opposite would happen. After all, the rate had been rising for many years, and there were plenty of reasons to believe that an event which worsened many of the normal indicia of suicide – economic loss, loneliness, mental health problems – would make things worse. But somehow, although the virus and the lockdown were awful for many, they weren't quite as unbearable as everyday life over the previous decade.

As you can show by considering how quickly permutations multiply up even in a squirrel farm like the one discussed in Chapter 4, the mathematics of complicated systems ensures that variety engineering at the unthinkably large scale has to be qualitative rather than quantitative. You can't say how much information a human being is taking in and reacting to at any given time, but you can easily observe the difference between a human being that is coping and one that is overloaded. That's my diagnosis of what led to the series of connected political eruptions between the financial crisis and the pandemic. The hypothesis set out as early as 1970 by Alvin Toffler in his book *Future Shock* turned out to be correct: the number of people who were no longer able to cope with the modern world reached a critical mass.

The clues were there. The economists Anne Case and Angus Deaton showed that the first two decades of the twenty-first century saw a rapid increase in 'deaths of despair': suicide, alcoholism and opiate addiction. Horribly but inevitably, Stafford Beer was not quite right to say that ignorance was 'the lethal variety attenuator'. There are even more fundamental ways of reacting to a situation that can't be dealt with or ignored.

Autonomy and the status syndrome

Two of the longest-running pieces of research in medicine are the Whitehall studies, the first of which began before Stafford Beer ever set foot in Chile. Whitehall I, from 1967 to 1977, recorded the general health outcomes of 17,500 male British civil servants; Whitehall II, ongoing since 1985, widened the remit to include women; both cohorts have been followed up and re-examined over the years.

Michael Marmot, who led the first study and initiated the second one, made an extremely interesting discovery. One of the strongest predictors of serious mortality outcomes – heart attacks, strokes, cancer – was the civil service grade that someone occupied. The very large sample size allows you to eliminate the effects of income, smoking, family history and a lot of other things and be sure that this is a separate effect. And social status (as represented by the civil service grade) is itself highly correlated with unhealthy behaviours, in a way that doesn't appear to be related to education, intelligence or any other variables that might be associated with self-control.

This 'social gradient' seems to be there in data from other countries, too. Marmot ended up concluding that the psychic feeling of being in control of your life is extremely important as a source of well-being, and that conversely, being out of control is physiologically harmful as well as emotionally intolerable.

At various points in this book, we've noted that you can tell when a cybernetic system is overloaded because it breaks down. Marmot's main conclusion from his research was that inequality in society was a major driver of public health risks, but it could be given a cybernetic interpretation too. The connection that he found looks like the result of a variety mismatch; people are, increasingly, unable to regulate the input from their immediate environment, and they correctly perceive this as a threat to health and life. That might be the deepest reason why managers create accountability sinks – to be accountable for something you can't change is to experience exactly the 'out of control' feeling that the Whitehall studies seem to suggest will kill you if you let it.

And what's true at one level of a system can be true of others. The breakdown in the economic and political system reflects the same imbalance that causes 'deaths of despair'. People are overloaded with information that they can't process; the world requires more decisions from them than they're capable of making, and the systems that are meant to shield them from that volatility have stopped doing the job.

The villain, unmasked?

And so the nature of the crisis is . . . it's not a crisis per se, it's the way that the system achieves long-term stability. The world isn't going to stop growing, so it will only get more complex.* That

---

* This isn't the same as saying that economic growth is going to continue for ever; measured GDP is one accounting system, and it might do all sorts depending on how things are organised. But the

means that systems have to be built that absorb the volatility and variety at the appropriate levels. The overall system is always looking for some organising principle of identity, which tells it what to ignore and how to balance the present and the future.

For fifty years, the free market played that role; the underlying guarantee was that all decisions would get taken care of, even if nobody made them. When that fell apart, the ultimate basis of the system – *el pueblo*, so to speak – sounded the alarm. Ever since then, we've been looking for a new organising principle.

I think this is what explains the common thread between MAGA, Brexit, the Five Star Movement, Hindutva and all the rest. The populist movements of the 2010s all promised a simpler world; they were, in the words of J. K. Galbraith, taking on the great anxiety of their people and addressing it. They were also promising to restore the broken communication channels – to make voices heard, to force the managerial class to listen.

But the same analysis tells us that they're fake solutions. You can't promise a simpler world – that's equivalent to claiming to be able to reverse the direction of time. And if you are promising to restore the broken communication channels, you need to say how. These channels used to be made up of layer upon layer of middle managers and civil servants. Not only would it be extremely costly to bring them back, it's not obvious that anyone would thank you for doing so. It's definitely not what the populists are proposing – there's no March for Bureaucracy, nobody's slogan is 'Red Tape Holds Us Together'.

---

world keeps having more people and things in it, more connections between those entities and so more bits of information to manage and regulate.

If any of this is correct, we ought to be able to say whether or not the system is capable of reaching stability again. And we ought to be able to diagnose it in Stafford Beer's terms. There are big mismatches in variety: areas of discrepancy between the volatility of the environment and the system's ability to handle it. Just identifying those areas ought to give some idea of where the big battles are to be fought.

I've often said in the past that if you don't make predictions, you'll never know what to be surprised by. Similarly, if you don't make recommendations, you won't know what to be disappointed by.

10

What Is to Be Done?

Should we all stand by complaining, and wait for
someone malevolent to take it over and enslave us? An
electronic mafia lurks around that corner.

Stafford Beer, *Designing Freedom*, 1974

Record producers have a saying: 'If you can hear the problem,
you can hear the solution.' Identifying what's wrong with a
mix is the same thing as identifying what needs to be changed.
Variety engineering is, unfortunately, not quite the same. Identifying which channels are broken, which systems are missing
or which translation mechanisms don't work is a significant
step forward, but there isn't a big desk in front of you with
a slider that will add more capacity wherever it's needed. As
the historian Alfred D. Chandler first identified, organisations
initially deal with imbalances in their ability to handle environmental variety by adding resources to the existing channels
– and when that is no longer possible, by reorganising. And
organisation isn't a scalar quantity – you can't just say how
much change you want.

You have to talk about specific changes. The viable system
model will help check whether any given proposal is capable of
working – it will tell you what to think about and whether it's
addressing the right problems. It will also help you think about

the other key parameter: the ability of the system to withstand and survive organisational change. But it won't generate ideas. We have to do that for ourselves.

There's one sense in which the recording engineers' maxim is more applicable, though. These are problems of human organisation, and the overwhelming majority of environmental variety is created by humans. The knowledge that something is an organisational problem means that the source of its complexity is the human beings in the organisation. That, in turn, means that the human beings in the organisation have sufficient variety and capability to match the complexity of the problem; the problem can't be bigger than them, because their own misdirected efforts created it. Only problems that aren't cybernetic can genuinely be insoluble.*

I don't have the capacity to make detailed designs of the society of the future – Stafford Beer himself didn't. But using his toolkit systematically lets us ask the right questions. What do we need to do, in order to stop the organisation of the industrial economy from causing excessive stress on the human population? If we can manage that, how can we begin to future-proof the system and make it adaptable to the fact of increasing complexity? But more fundamentally, how can we begin to get rid of the theoretical and ideological blind spots that got us into this mess in the first place?

Starting with that last question, it's time to find out what happened to Stafford Beer after Chile.

---

* One implication of this proposition is that it gives us a rigorous definition of what would constitute environmental Armageddon – it's the singularity point at which the problem of climate change ceases to be soluble purely by changing human behaviour and organisation.

A meeting in Cwarel Isaf

The year is 1975. Brian Eno is up to his ankles in mud, being enthusiastically greeted by two large dogs. He has come to an isolated cottage in Wales to meet Stafford Beer for the second time. They talk for hours about algorithms, self-organising music and the future of democratic society, until Eno practically collapses from hunger. When Beer tells Eno that he wants to pass on the torch of cybernetics, the musician replies, 'I'm very flattered that you think that, but I don't see how I can accept it without giving up the work that I do now – and I'd have to think very hard about that.'

It might sound like a scene from a psychedelic movie, but this conversation actually happened. And although booze had clearly been consumed, Beer's offer appears to have been made seriously. The first meeting had happened earlier that year in London, after Eno had sent Beer an essay he'd written on 'Generating and Organising Variety in the Arts', an adaptation of cybernetic principles to musical performance. Beer's life had not settled down after his time in Chile. He was recently divorced and working on the expanded second edition of *Brain of the Firm*, which included his assessment of the Allende experience. Eno, on the other hand, had been interested in generative art and cybernetic themes since art school. On receiving a copy of the first edition of *Brain of the Firm* from his mother-in-law, he had devoured it.

Eno was at a different point in his career; he had left the art pop band Roxy Music two years earlier and had by this point released two solo albums. In 1975, he was working on the first recordings of *Discreet Music* (the name 'ambient' only came along later) and *Another Green World*. His studio techniques were already at an advanced stage of development; the Genesis album *The Lamb Lies Down on Broadway*, released in 1974, credits him with 'Enossification'. Stafford Beer wasn't looking for a missionary for management cybernetics in the pop world; he was looking for someone to carry on his academic work. In context, it's not hard to see why this didn't happen.

Beer never solved the problem of who he might pass the torch to; he seemed to be looking for creative people who wanted to change the world, but people like that tended to have things of their own going on. He held many visiting professorships, but never took on the kind of stable full-time academic job that would have ensured he had PhD students to carry on his legacy. He did, however, recruit one more evangelist in the music world. A couple of years after the meeting at the Welsh cottage, Eno was helping David Bowie to produce some of the most influential music of the twentieth century in Berlin; according to others present he was constantly drawing diagrams and inventing processes and systems to make things happen. Bowie was quick to get cybernetics and the message stuck – as late as 1995, he was mentioning *Brain of the Firm* when asked about his top ten books by the *Daily Mirror*.

Endless possibilities

Thirty-one years after that meeting with Stafford Beer, Brian Eno produced an artwork called *77 Million Paintings*. I saw it in an exhibition at the Barbican in London in 2006, alongside a festival dedicated to the composer Steve Reich.

The exhibition was thematically linked to Reich's work; he's a minimalist composer, and a lot of his early work in particular was built around layering and repeating bits of noise and music, creating new and strange sound events. Eno's work was a piece of what he calls 'generative art'. LCD screens were within picture frames, and behind them a computer was running a piece of software that generated the pictures. The software took a few visual elements and recombined them according to a small set of rules – it might layer them across each other, stretch and rotate them or change the colour palette. As we've seen in previous chapters, the variety in such simple rules is multiplied; from a relatively compact piece of code, it was possible to calculate that the rules would over time generate 77 million distinct images.

These are decision-making systems. You'd have to stretch the definition of 'artificial intelligence' to absurd lengths to cover them, but they make decisions about what images to generate; they progress and they have a purpose. You could even say that they're embedded into a recursive system for higher-level decision-making; if one of them started showing loads of swastikas, for example, somebody would probably step in and do something. But their purposiveness doesn't consist in maximising anything; they're just exploring the possibilities that are open to them.

That particular piece of art didn't respond to its environment, but it's not much of an imaginative leap to think about one which could do so. In fact, interactive generative art has been around from the earliest days of cybernetics. Gordon Pask, the guy with whom Stafford Beer tried to grow an artificial ear, attempted to commercialise the MusiColour lighting system, which would adapt a theatrical show to the way the audience appeared to be dancing. Many mobile phone apps exist that act as a modern handheld version of MusiColour, recombining music and shapes in response to touches and swipes – Eno has produced a few, as have plenty of other artists and developers. These are systems with the key cybernetic property of 'ultrastability'; they are able to find a steady state in response to conditions which their designers didn't anticipate.

But they don't work by defining an objective function and seeking to maximise it. This is the paradigm shift that might be required – that organisations and systems can be like people, having purposes without a single goal. An artist doesn't have a successful career by maximising their art; they do it by repeatedly producing work that they are proud of.

That's what the world could look like if we got rid of the blind spots. Businesses ought to be like artists, not paperclip maximisers. The economic concept of optimisation, and the institutions of management and government which enforce its use, effectively act as a brutal information-reducing filter. By taking away the pressure to maximise a single metric (and therefore to throw away information that doesn't relate to it), organisations could apply their decision-making capabilities much more effectively. They could innovate more, design more sustainable solutions and build less adversarial, longer-term relationships with their people.

Restoring the polarity

Oddly enough, this is quite close to another vision of the economy that got forgotten in the 1970s. I mentioned in Chapter 8 that the only one of Milton Friedman's contemporaries who had anything like the same public profile as an economist was the Canadian J. K. Galbraith. This was arguably underselling things – for a lot of that time, and outside the USA, Galbraith was the bigger star.* And J. K. Galbraith saw the role of corporations very differently to Friedman.

Galbraith wasn't a mathematical economist; he didn't write

---

* Milton Friedman won his Nobel Prize in 1976, but his place in popular culture was cemented in 1980 by a very popular TV series broadcast on public television called *Free to Choose*. This had been commissioned specifically as a conservative 'right to reply' to Galbraith's *Age of Uncertainty*, which had been broadcast three years earlier.

down systems of equations and prove that they had optimised solutions. That's one of the reasons why his economics has been somewhat left behind by the modern profession. But he thought a lot about planning and information, and recognised that the big fact of the post-war economy was the growing importance of large organisations. He was also one of the few economists to seriously address the growing importance of the professional and managerial class. He generally referred to it as the 'technostructure' – a complex category spanning business, government, military and media, working in the interest of maintaining overall stability.

It's quite startling to go back and read Galbraith side-by-side with a few issues of the *JACF*, because it's an utterly different view of how corporations behave. He didn't believe in the financial market as a source of judgement and incentives – he thought that executives' main priority was to handle uncertainty. He actually has a theory of advertising that makes intuitive sense: it's simply another part of planning. Given the amount of work and investment that goes into launching a new product, it would be somewhat absurd for companies to put all their effort into the supply side, and not to make any attempt at all at planning the demand.

The Galbraithian ideal economy is one in which the technostructure prevails; the business cycle is managed by managers, who maintain stability by communicating with one another, and keep control of the productive resources of society simply by showing up to do a job that nobody else wants to bother with. He is in general suspicious of 'competition' when it shows up – absolute monopolies, in his view, are likely to be exploitative, but cosy arrangements to fix prices are more likely to be a good thing. They generate profits which can be channelled into investment and innovation, they bring stability to the overall price and wage bargains, and they encourage businesses to compete by improving quality.

This is only one of the models which got lost in the mathematical revolution; a Nobel Laureate called Herbert Simon also proposed a 'behavioural theory of the firm', in which profits were not the goal to be maximised, but instead a constraint on behaviour – managers were assumed to do just enough to keep investors happy, then spend the rest of their time and effort trying to do new things they could be proud of. There are a number of models, most of them ignored for decades, in which the corporate sector provides a stabilising function, insuring the working class against fluctuations in the business cycle, rather than expecting them to soak up the volatility.

The intriguing thing is that Simon and Galbraith didn't write polemics to the effect that this was how corporations should behave – they just described what was in front of them at the time. Before Milton Friedman's essay, lots of people assumed that this was just naturally the way things would tend. Without the Friedman doctrine, without very great re-engineering of the systems of corporate finance, the industrial economy might have just gone on and developed into a technostructure.

Maybe they were right? It would certainly be good if they were, because that might indicate a much easier path to defuse the immediate source of crisis. If the problem with the modern corporation is the result of the capitalist counter-revolution against the managerial class, we just need to change the terms of the battle.

The new nature of the corporation

As far as I can see, the first step towards the possibility of any improvement has to involve doing something about private equity. There are other problems with modern management practices and pathologies of the financial system, but this is the big one. It causes a huge number of problems, it normalises bad practice and it doesn't provide proportionate benefits. Dismantling the leveraged buyout industry would get rid of an overhanging threat across the entire managerial class; it would open up a huge space for different models of corporate governance.

But how would you go about doing that? If someone thinks that they could run a company better than the existing management, they're allowed to spend their money buying it; preventing that from happening would involve a great disruption to the system. You can't stop the management of a company from taking on debt, either. A lot of the time, taking on debt – even risking bankruptcy – is the correct and necessary thing for a company to do. As soon as you start trying to design regulatory regimes to prevent 'excessive' corporate borrowing, or distinguishing between companies on the basis of their owners, you quickly start to realise that this is a case where the law of requisite variety is telling you the project is doomed* – the variety of financial situations that a company can be in is much greater than anything that could be written into a rule.

The same principles of variety engineering show where the solution has to be. We know that the bad thing about debt in this context is that it swamps all other signals and distorts

---

* Buried here, in this footnote towards the end of the book, is probably the most useful and valuable piece of advice in all its pages: checking whether Ashby's law of requisite variety has been respected is a great way of spotting a doomed project. It will even tell you if the project is going to fail because of insufficient resources or fail because it's an impossible thing to achieve.

decision-making, so we need to look at where feedback is amplified, attenuated or removed. In context it's pretty clear. The principle of limited liability acts as an accountability sink for private equity investors. It allows them to take on debt, and then extract cash in the knowledge that if things go badly, this information can't travel upstream to affect the entity that took the decision.

In the language of Beer's model, the principle of limited liability is part of a 'resource bargain'. As we've already mentioned, this is the fundamental unit of governance for a viable system. It's the means by which a subsystem is granted decision-making autonomy, so long as its decisions don't affect broader systemic stability. Limited liability was introduced for joint-stock corporations in the USA in the 1820s, part of a bargain that would allow access to a much larger pool of savings in order to finance the industrialisation of a new continent. It came to England thirty years later, and was eventually adopted all over the world. But it's a bargain, not a principle of natural justice, and it needs to be reviewed if it's causing trouble.

Former US presidential candidate Elizabeth Warren had an eight-part plan for reforming the private equity industry,* but from a cybernetic point of view only one of these points was needed: any entity taking control of an operating company should have to guarantee its debts. You might also want to impose some rule on the payment of exceptional dividends, but I doubt it would be necessary. Once you take away the big accountability sink – and make it possible for a bad deal to

---

* Early on in the LBO boom, in 1985, Paul Volcker of the Federal Reserve also had a proposal, based on directly regulating the amount of leverage allowed to be used by shell companies in acquisitions. It has been suggested that this was among the reasons why the Reagan administration replaced him with Alan Greenspan in 1987.

affect investors as well as the little people – most of the attraction of using a company's debt-servicing capacity to 'buy it with its own money' is gone. You can keep the good part of the resource bargain for limited liability, but concentrate it on its original purpose – allowing small, non-controlling investors to contribute to the purchase of industrial capital. That's where the major benefit of the arrangement is, so there's no need to allow further levels of limited liability.

Viability over maximisation

I think this might be all the change needed to destroy Friedmanism. Two decades into the twenty-first century, it's easy to think of the crazy ideology of profit maximisation as inescapable. When we think of corporations as artificial intelligences, we worry they have developed a monomania, an addiction to a particular kind of stimulus, and that they are destroying their own lives, the lives of others and even the material conditions of their existence, in order to satisfy their craving for profits.

But that is straining at the edges of what might be justifiable as a metaphor. Corporations are decision-making systems, not 'intelligences'. They have homeostatic forces which aim to maintain their equilibrium, and higher-order decision-making systems which mean they are able to reorganise themselves in order to respond to shocks beyond the scope of anything anticipated when they were designed. To attribute motivation to them is to make assumptions about the internal workings of the black box – the original intellectual sin of cybernetic analysis. The only reason we might think that the purpose of an oil company is to destroy human life on the planet Earth is POSIWID – that's what they do.

They do this because they have to. The decision-making

system of a modern corporation has a particular dysfunction: one of its signals has been so amplified that it drowns out the others. The 'profit motive' isn't something that can be ascribed to corporations – they don't have motives. What they have is an imbalance between the two key higher functions – here-and-now versus there-and-then. They aren't capable of responding to signals from the long-term planning and intelligence function, because the short-term planning function has to operate under the constraints of the financial market disciplinary system. Either a corporation has a survival condition based on needing to make a monthly interest bill, or there's an implicit threat from the financial environment that if it fails to behave in a particular way, it will be taken over by an outside entity.

If you take away that pressure, it's quite likely that the natural equilibrium of corporate decision-making systems will be less hostile to human life. Viable systems fundamentally seek stability, not maximisation. In the era of capitalism that J. K. Galbraith and Herbert Simon described, the giants of corporate America were content with being big and important; they didn't feel the need to get every last bit of cash that could possibly be extracted.

And although there were undoubtedly pockets of laziness and corruption, it wasn't all bad. As well as corporate jets, they funded lavish research labs. As well as making agreements to stitch up markets, they tried to compete on quality and innovation, not just price. If you compare the allegedly sclerotic era of Galbraith's 'technostructure', before the leveraged buyout boom, with the modern era of dynamism and pressure, you might be surprised to see which one performed better – even in terms of economic growth.

In other words, if we believe that the greatest problem facing humanity is that the artificial intelligences which rule our daily lives have gone mad, then this is a cybernetic problem – and it might have a surprisingly simple cybernetic solution. It is hard to hear the conscience of corporations above the noise of debt and 'fiduciary duty', but it's there. If you quieten the shrieking klaxon of the debt burden, you'll change their information environment. On any given day, managers spend a lot more time talking to their customers and employees than they do to investors; if they were able to pay attention to what they heard, that would be much healthier for their decision-making.

Our colleagues the economists

That would only be the first step, though. There is a lot more to global governance than corporate management, and as long as the ideology of economics maintains its dominant position, there is always a considerable danger of the Friedman doctrine rising back up from the dead. If the highest-level decision-making mechanisms of the world are to be viable systems, they need a philosophy which can balance present against future and create self-identity. We saw the reasons for that in Chapter 5. And it's possible to deduce from the rest of the model that this philosophy *cannot* look much like mainstream economics, in its current state.

The reason why economics can't be the governing philosophy is based in the same underlying fact that motivates the AI researchers' fears of paperclip machines:

Any system which is set up to maximise a single objective has the potential to go bonkers.

It follows from the mathematics of constrained optimisation, combined with the basic laws of cybernetics. Setting up a maximising system involves defining an objective function, and throwing away all the other information. Sooner or later, the environment is going to change, and something which isn't in the information set any more is going to lead the system into destruction.

Consequently:

> Every decision-making system set up as a maximiser needs
> to have a higher-level system watching over it.

There needs to be a red handle to pull, a way for the decided-upon to indicate intolerability. When you hand over responsibility to an algorithm – either a literal one, or a part-human system with instructions to maximise something – part of the resource bargain has to involve making available the oversight capability, to make sure that the maximisation doesn't get out of control.

And that in turn implies a quite definite conclusion for the design of real-world systems:

> You can't have the economists in charge, not in the way
> they currently are.

If every maximising system has to have a higher-level system governing it (to make sure it doesn't go bonkers), then that logically implies that the top level of any decision-making system that's meant to operate autonomously can't be a maximiser. And so, the governing philosophy of the overall economic system can't be based on the constrained optimisation methodology that's currently dominant in the subject of economics. Otherwise there's a risk that the system will go bonkers, and that it will start pursuing maximising objectives, oblivious to the danger that it's on course for making human life impossible. Like it actually has done.

This sounds like a bitter pill for the economists. It's nothing like as bad as it sounds, though. For one thing, it's not as if the toolkit of optimisation needs to be thrown away completely. As we said before, if you have some inputs and you want some outputs, then you want to get the most outputs for your inputs, and that's what economics is all about. Providing a governing ideology and philosophy isn't the only thing that makes a science worth doing – John Maynard Keynes once said that economists could consider their discipline a success when they were regarded as useful and competent technicians, like dentists.

And for another thing, economics contains multitudes. The turn towards maximisation and mathematics is relatively recent; J. K. Galbraith and Herbert Simon are mentioned in this chapter, but there are lots of other traditions in which this runaway maximising behaviour isn't the central dogma. If they want to remain the queen of social sciences, economists just have to change their priorities. Like any other viable system, economics is capable of restructuring itself in response to environmental necessity.

Our friends the robots

It would be a great thing if the economy could be made to stop channelling stress and volatility into people's lives. But although that would go a long way towards reducing the immediate threat, it's only a first step. We still have the basic problem that the system requires the creation of accountability sinks in order to function, that the general population are not happy with this, and that the only ways they have of expressing their discontent seem to be highly destructive of the system itself.

In Chile, Stafford Beer's dream was to create an engineered communication channel of this kind – every household would have a dial they could twist to indicate their general level of satisfaction with the system, which would trigger urgent research and question-asking from the top if the aggregate signal dropped below a target level. But this never happened – Beer's teenage son built a prototype in the family shed, but that was it.

Today we have social media. In principle, they could play a similar role in conveying information from the grass roots into the heart of our decision-making systems. But that would be difficult. It would involve a very significant change in the amount of information that those systems had to take in. And one thing we know from Alfred D. Chandler's history of management is that a big change in complexity will usually require an equally significant reorganisation. So far, we've done little to meet that challenge. All we've done is use electronic communication to create a more efficient form of the old-fashioned technology called 'shouting'.

We could do better. We even have the technology to do so. The social media system is a source of excess variety, which our governance systems can't currently handle. But during the period in which these technologies have been expanding, we've also been inventing incredibly powerful new means of variety *amplification*. A modern artificial intelligence system – a transformer recurrent neural network* can take a large block of text

---

* At the time of writing, ChatGPT is the market leader, but that might have changed by the time you read this. This is the 'Silicon Valley near future' tense again – when I started writing this book, I was only aware of these things because a friend of mine had quit his hedge

and summarise it quickly. It can also expand a short instruction into a longer explanation. It's practically designed for facilitating two-way communication between a mass audience and a smaller decision-making system. It would really be a generational shame if we ended up once more just using it to make our existing structures work faster – like bringing back Shakespeare, Machiavelli and Napoleon and setting them to work designing tax forms.

But in order to do that, we are going to need to loosen up when it comes to dealing with accountability sinks. In the future, as well as being made by unaccountable black box systems, important decisions are going to be made by actual robots. This isn't a choice we're faced with, unfortunately. It's just a consequence of things getting more complicated, and passing the threshold at which they have to be analysed as a whole. We cannot afford the luxury of explainability; we can't keep on demanding that an identifiable human being is available to blame when things go wrong.

Management cybernetics doesn't give any clues as to how such a profound social change might be achieved, unfortunately. I could make something up, but it would feel like a shabby way to treat you after we've come so far together.* My only guess is that it might be that what's really intolerable about unaccountability is the broken feedback link, and that if we can solve the problem of communicating with the system – pay more attention to the 'red-handle alert' mechanisms that indicate an unbearable outcome – people might not be so furious about

---

fund job to work on them. By the time I finished it, they were on the evening news.
* Particularly Chapter 5, which I promise was a lot harder to write than it was to read.

the death of personal responsibility. In general, people in my experience are a lot less angry about everything when they feel like they're being listened to.

In the future, 'I blame the system' is something we will have to get used to saying, and meaning it literally.

Further Reading

I should probably start with a warning – cybernetics can be an expensive hobby. Lots of the books mentioned below were originally published by academic presses, and so they tend to sell at prices which even a university librarian might gulp at. I can't, in all conscience, encourage anybody to go out and start acquiring the complete works of Stafford Beer at list price. However, it seems that the 1980s generation of management consultants are gradually retiring and downsizing their libraries, so they often show up in charity shops and second-hand bookstores at reasonable prices. That's how I got most of mine. Consequently, some of the references below might be to out-of-print editions, or ones where the original publisher doesn't exist in that form any more.

Another warning might be that some of them have a lot of maths, and many of the rest have lengthy verbal descriptions of mathematical models which aren't always much less off-putting. For the most part, you can skip over the equations, but in my view if you're going to seriously study this subject, start with *An Introduction to Cybernetics* by W. Ross Ashby (Chapman & Hall, 1976), which was meant to be a basic text for biomedical researchers and which is much more clearly written than any of the others, with only high-school mathematics required.

If you want to read more about cybernetics, but not so seriously, *The Cybernetic Brain: Sketches of Another Future* (University of Chicago Press, 2011) by Andrew Pickering is great fun; you will realise that I only scratched the surface when it comes to the eccentricity and genius of the British cyberneticians. It's one of the only books I'd wholeheartedly recommend paying full academic price for if you can't get it any other way. James Gleick's *The Information: A History, a Theory, a Flood* (Fourth Estate, 2012) is also an acknowledged classic, and tells the story from the other side of the Atlantic – what the telecoms engineers and computer programmers did at and after the Macy conferences.

In terms of Beer's own works, the big two are *Brain of the Firm* (John Wiley, 1972, but if possible get the second edition from 1981 which has a lot of updated material) and *The Heart of Enterprise* (John Wiley, 1979). Those are the correct names, by the way; if you don't want to look like a newbie, never say '*The Brain of the Firm*' or '*Heart of Enterprise*'. Of the two, *Heart* is a lot more user-friendly while *Brain* is the really essential rigorous text of the viable system model. Don't allow yourself to get stuck on page 47 of *Brain*; the 'anastomic reticulum' diagram in Figure 9 has confused everyone I know who has read it and it's much easier to just work out the point it makes with a pencil and paper. Beer's own textbook is *Diagnosing the System for Organizations* (John Wiley, 1985) and it's a very good introduction with all the maths replaced by diagrams if you like that sort of thing.

There's a lot of good stuff in later Beer too, although it's mainly on wider issues than those discussed in this book. *Platform for Change* (John Wiley, 1975) gives his views on social and macroeconomic questions while *Designing Freedom* (John Wiley, 1975) is a shorter version of the same theses, compiled from his CBC Massey Lectures. Two collections of shorter pieces and lectures are *How Many Grapes Went into the Wine*, edited by Roger Harnden and Allenna Leonard (John Wiley, 1994) and *Think Before You Think*, edited by David Whittaker (Wavestone Press, 2009). These are probably the best place to start, because you can dip into them rather than needing to clear time to hold the whole structure in your head, and you'll almost always find something amusing and insightful when you do. I also greatly enjoyed *Decision and Control* (John Wiley, 1966) and *Management Science: The Business Use of Operations Research* (Aldus Books, 1968), but these are both more didactic management textbooks and predate the viable systems model, so they should probably be considered 'lesser works'. *Pebbles to Computers: The Thread* (co-authored with David Suzuki and with gorgeous photos by Hans Blohm, OUP, 1986) is great to lay out on your coffee table.

In terms of books by other people about Beer, I'd thoroughly recommend *Stafford Beer: The Father of Management Cybernetics* by Vanilla Beer and Allenna Leonard (independently published, 2019). It's co-authored by his daughter and long-term partner, and has a lovely biography in the form of a graphic novel combined with an incredibly useful glossary of cybernetic terms. The latter is particularly essential because you often come across quite specialised language in these books and it's very helpful to get a steer from someone who knows *exactly* what he's talking about. David Whittaker's *Stafford Beer: A Personal Memoir* (Wavestone Press, 2003) has some nice details in it and a very good interview with Brian Eno.

A more technical guidebook is *The Viable System Model* (John Wiley, 1989), a collection of applications of and essays on Beer's work, edited by Raúl Espejo (one of the original Cybersyn team) and Roger Harnden. Robert Flood and

Michael Jackson's *Creative Problem Solving* (John Wiley, 1991) is probably the one to read if you really fancy having a go at applying management cybernetics to a real-world consulting assignment, although many friends swear by *The Fractal Organization* by Patrick Hoverstadt (John Wiley, 2011). Of course, Eden Medina's *Cybernetic Revolutionaries* (MIT Press, 2011) is the definitive account of what really happened in Chile.

Norbert Wiener's *Cybernetics* (MIT Press, 1948) is hard to recommend and in all honesty I don't believe that the majority of people who bought it got past the first chapter. It has some of the foundational mathematics of information theory in it, and it's a cultural artefact of its time, but *The Human Use of Human Beings* (Riverside Press, 1950) is a lot easier to read. His biography, *Dark Hero of the Information Age* by Flo Conway and Jim Siegelman (Perseus Books, 2005), is the most user-friendly way of getting the information, in my opinion. With respect to managerialism, though, James Burnham's *The Managerial Revolution* (John Day, 1941) is worth going to the primary source – it's clearly written and has aged rather well. The same is true of *The New Industrial State* by J. K. Galbraith (Houghton Mifflin, 1967), and of *Strategy and Structure* (MIT Press, 1969) by Alfred D. Chandler.

Free to Choose by Milton and Rose Friedman (Harcourt, 1980) is the book mentioned in the text and it's a good read. Nicholas Wapshott's *Samuelson/Friedman: The Battle Over the Free Market* (W. W. Norton, 2021) gives the flavour of just how important and influential Milton Friedman was, through a review of his constant sparring matches with fellow Nobel Laureate Paul Samuelson in their respective *Newsweek* columns. Philip Mirowski and Dieter Plehwe's *The Road From Mont Pèlerin: The Making of the Neoliberal Thought Collective* (Harvard University Press, 2009) fills in a lot more background of how neoliberalism prepared its ideology when it was getting ready for the leveraged buyout boom to take off.

Mirowski's *More Heat than Light* (Cambridge University Press, 2011) and *Machine Dreams* (Cambridge University Press, 2002), along with E. Roy Weintraub's *How Economics Became a Mathematical Science* (Duke University Press, 2002), give more background on the development of 'mainstream' economics than it was possible to cram into Chapter 6. Among many works detailing the state of management science, I'd recommend *Management Studies in Crisis: Fraud, Deception and Meaningless Research* by Dennis Tourish (Cambridge University Press, 2019) and *Nothing Succeeds Like Failure: The Sad History of American Business Schools* by Steven Conn (Cornell University Press, 2019). It is a bit of an indictment of things that H. Thomas Johnson and Robert S. Kaplan's *Relevance Lost: The Rise and Fall of Management Accounting* (Harvard Business School Press, 1987) is still itself every bit as relevant today as when it was published.

There is no shortage, to say the least, of books detailing the foibles of the management consulting industry – *The Management Myth* by Matthew Stewart (W. W. Norton, 2009) has the advantage of having been written by someone who worked in the industry and was prepared to tell tales. *The Big Con* by Mariana Mazzucato and Rosie Collington (Allen Lane, 2023) came out just after I had finished writing about the decerebration of the public sector and covers that topic thoroughly and excitingly.

You really would need an academic library to find 'Reasoned Argument and Social Change' (National Communication Association, 2011), but if you happen to be in one, it is worth looking up to find Frans H. Eemeren, Bart Garssen and Jean H. M. Wagermans' masterful analysis of the KLM squirrel apology press release. *Flying Blind: The 737 MAX Tragedy and the Fall of Boeing* by Peter Robison (Penguin Business, 2021) is also an enthralling history of a massive accountability sink. I learned a huge amount from Gill Kernick's *Catastrophe and Systemic Change: Learning From Grenfell* (London Publishing Partnership, 2021) about another case study in which interlocking systems combined to create a tragedy that couldn't systematically be pinned on any single organisation, let alone an individual.

Brian Alexander's *Glass House: The 1% Economy and the Shattering of the All-American Town* (St Martin's Press, 2017) is the best of the 'private equity hometown memoir' genre mentioned in Chapter 8; Gretchen Morgenson and Joshua Rosner have covered the ongoing economic disaster in *These Are the Plunderers* (Simon & Schuster, 2023). Moe Tkacik's journalism at the American Economic Liberties Project continued to shock me long after I thought I had seen it all from the private equity industry. David Gelles, in *The Man Who Broke Capitalism* (Simon & Schuster, 2022), does the same thing with respect to Jack Welch. If you're only familiar with the modern form of capitalism and management, it's very worth reading *As I See It* (Prentice-Hall, 1976), the autobiography of J. Paul Getty, to see that things really did used to be very different.

A Citizen's Guide to Artificial Intelligence (MIT Press, 2021) by John Zerilli, John Danaher, James Maclaurin, Colin Gavaghan, Alistair Knott, Joy Liddicoat and Merel Noorman is exactly what it says it is and covers the important issues clearly, written with knowledge of something close to the current state of technology. John Searle's *Minds, Brains and Programs* (Cambridge University Press, 1980) has the original presentation of the 'Chinese room' argument, while *Consciousness Explained* (Little, Brown, 1991) by Daniel C. Dennett gives the opposing view. Ben Kuipers' paper on corporations as artificial intelligences was presented at the MIT conference on *Collective Intelligence* in 2012 and can be downloaded from his website, while Charlie Stross coined the 'very old, very slow AI' phrase in his address to the 34th Chaos Communication Congress; it's also on his website.

I will confess that I didn't get much out of *Practical Iron-making* (United Steel, 1959) by G. D. Elliot and J. A. Bond.

Acknowledgements

This book was a while in the making, and involved changing my mind on key concepts several times after talking to other people. Among those who might recognise some of the points they made to me over the years are Andrea Filtri, Mark Blyth, Chris Bertram, Trevor Petch, Henry Farrell, Kieran Healey, Dan Hardie, Steve Teles, Chris Brooke, Doug Henwood, Ben Kuipers, Charlie Stross, Alex Harrowell, Indi Samarajiva, Jo Michell, Josh Loftius, Jamie Galbraith, Alex Williams, Stian Westlake, Sam Freeman, Maria Farrell, Carolyn Sissoko and Daniela Gabor. I dearly wish Sebastian Nokes and Kelley Walker were still here to do the same. Every one of them has a perpetual licence to say that they came up with all the good bits and express confusion at how I got all the rest of it so mixed up.

Michael Pollak also belongs in the list above, but he made so many useful comments on the actual book that I have to start a new paragraph to mention him specifically. He challenged a number of the key concepts and forced me to get rid of a number of digressions and self-indulgences. He has promised to tell me what his *real* disagreements with the thesis are when it's published, and I look forward to those! Tim Atkins also made really useful comments. Thanks also to Brian Eno for spending some of his presumably extremely rare spare moments on an email exchange confirming the details of what he did and said.

Acknowledgements

Ed Lake commissioned the book, and his strength of will in rejecting numerous proposed outlines and telling me to get it right has undoubtedly made the thing much better. Nick Humphrey continued Profile Books' insistence on quality control and standards, and has helped hugely to turn a list of hobby-horses and semi-related anecdotes into a coherent argument. My agent, Sophie Hicks, kept the process moving with admirable efficiency. I also benefited hugely from working on another book at the same time – 'The Brompton', co-authored with Will Butler-Adams, who demonstrated to me that an intelligent engineer and CEO is likely to independently invent a lot of the solutions that the cyberneticians came up with.

Darren Sharma was, as he always is, a source of constant optimism and good vibes, which helped a great deal during the periods when I was starting to believe I had taken on a subject too large for my brain to contain without damage. My darling Tess had to put up with me during these moments, so for this and so much else she has my perpetual thanks.

Index

Index

Index

Index

About the Author

Dan Davies is a former Bank of England economist and investment bank analyst. As a journalist he has tackled the LIBOR and FX scandals, the collapse of Anglo Irish Bank and the Swiss Nazi gold scandal. He has written for the *Financial Times* and the *New Yorker*, and is the author of *Lying For Money*.

Made in United States
Cleveland, OH
26 December 2025

30159236R00167